Essays in
General Relativity

ESSAYS IN
GENERAL RELATIVITY

A Festschrift for Abraham Taub

Edited by Frank J. Tipler

Center for Theoretical Physics
University of Texas at Austin
Austin, Texas

 1980

ACADEMIC PRESS
A Subsidiary of Harcourt Brace Jovanovich, Publishers
New York   London   Toronto   Sydney   San Francisco

6448-1840

PHYSICS

ACADEMIC PRESS, INC.
111 Fifth Avenue, New York, New York 10003

United Kingdom Edition published by
ACADEMIC PRESS, INC. (LONDON) LTD.
24/28 Oval Road, London NW1 7DX

Library of Congress Cataloging in Publication Data
Main entry under title:

Essays in general relativity.

"Publication list of A. H. Taub" : p. 233
Includes bibliographical references.
   1. General relativity (Physics)--Addresses, essays,
lectures. 2. Taub, Abraham Haskel, Date --Addresses,
essays, lectures. I. Taub, Abraham Haskel, Date
II. Tipler, Frank J.
QC173.6.E83   530.1'1   80-517
ISBN 0-12-691380-3

PRINTED IN THE UNITED STATES OF AMERICA

80 81 82 83   9 8 7 6 5 4 3 2 1

Abraham H. Taub

To Abe Taub, The Universe Man

He masters the biggest equation
of the biggest shock wave
of the biggest bang that ever was.
Hydrodynamics and entropy,
they phase him not;
they are his meat and drink.
He creates universes right and left.
To lesser men he leaves the job
of distinguishing between a left-handed
and a right-handed Taub universe.
May he keep waves and shocks,
stars and universes,
forever in happy pursuit of one another.
And may we continue to enjoy
the fruits of his work
for many more years.
Three cheers for Abe Taub,
The Universe Man!

John Archibald Wheeler
High Island
27 June 1978

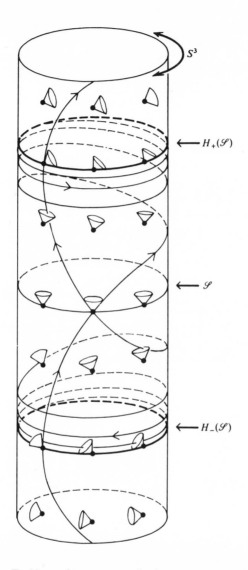

The Taub Universe. Taub's most important contribution to general relativity was his discovery of the Taub universe [*Annals of Mathematics* **53**, 472 (1951)]. This model has been a major source of examples of strange behavior that can occur in strong gravitational fields. For example, the above figure, which is a drawing of the causal structure of the Taub universe, pictures null geodesics that enter but never leave a compact set. These geodesics wind around the Taub universe an infinite number of times in a finite affine parameter length of the geodesics. Other examples of peculiar behavior in a Taub universe are discussed in the essays of J. A. Wheeler and L. C. Shepley. [Figure from "Structure of spacetime," by R. Penrose, in *Battelle Rencontres,* edited by C. M. DeWitt and J. A. Wheeler (Benjamin, New York, 1968); reprinted by permission.]

Contents

9 Locally Isotrophic Space–Time Nonnull Homogeneous Hypersurfaces

M. A. H. MacCallum

10 The Gravitational Waves That Bathe the Earth: Upper Limits Based on Theorists' Cherished Beliefs

Mark Zimmermann and Kip S. Thorne

11 General Relativistic Hydrodynamics: The Comoving, Eulerian, and Velocity Potential Formalisms

Larry Smarr, Clifford Taubes, and James R. Wilson

List of Contributors

Numbers in parentheses indicate the pages on which the authors'contributions begin.

Dieter R. Brill (13), Department of Physics and Astronomy, University of Maryland, College Park, Maryland 20742

Arthur E. Fischer (79), Department of Mathematics, University of California at Santa Cruz, Santa Cruz, California 95064

Robert T. Jantzen (97), Department of Physics and Astronomy, University of North Carolina at Chapel Hill, Chapel Hill, North Carolina 27514

E. P. T. Liang (191), Institute of Theoretical Physics, Department of Physics, Stanford University, Stanford, California, 94305

André Lichnerowicz (203), Collège de France, 75014 Paris, France

Lee Lindblom (13), Department of Physics, University of California at Santa Barbara, Santa Barbara, California 93106

M. A. H. MacCallum (121), Department of Applied Mathematics, Queen Mary College, London E1 4NS, England

Jerrold E. Marsden (79), Department of Mathematics, University of California at Berkeley, Berkeley, California 94720

Charles W. Misner (221), Department of Physics and Astronomy, University of Maryland, College Park, Maryland 20742

Vincent Moncrief (79), Department of Physics, Yale University, New Haven, Connecticut 06520

A. Papapetrou (217), Laboratoire de Physique Théorique, Institut Henri Poincaré, 75231 Paris, France

Roger Penrose (1), Mathematical Institute, Oxford University, Oxford, England OX1 3LB and Department of Mathematics, University of California, Berkeley, California

*Tsvi Piran** (185), Center for Relativity, University of Texas at Austin, Austin, Texas 78712

L. C. Shepley (71), Department of Physics, University of Texas at Austin, Austin, Texas 78712

*Present Address: Institute for Advanced Study, Princeton University, Princeton, New Jersey 08540.

*Larry Smarr** (157), Departments of Astronomy and Physics, Harvard University, Cambridge, Massachusetts 02138

Clifford Taubes (157), Department of Physics, Harvard University, Cambridge, Massachusetts 02138

Kip S. Thorne (139), California Institute of Technology, Pasadena, California 91125

Frank J. Tipler (21), Center for Theoretical Physics, University of Texas at Austin, Austin, Texas 78712

John Archibald Wheeler (59), Center for Theoretical Physics, Department of Physics, University of Texas at Austin, Austin, Texas 78712

James R. Wilson (157), Lawrence Livermore Laboratory, University of California, Livermore, California 94550

James W. York, Jr. (39), Department of Physics and Astronomy, University of North Carolina at Chapel Hill, Chapel Hill, North Carolina 27514

Mark Zimmermann† (139), W. K. Kellogg Radiation Laboratory, California Institute of Technology, Pasadena, California 91125

*Present Address: Departments of Astronomy and Physics, University of Illinois, Urbana, Illinois 61801.

†Present Address: IDA/SED, 400 Army-Navy Drive, Arlington, Virginia 22202.

Preface

This volume is a collection of essays to honor Professor Abraham H. Taub on the occasion of his retirement from the mathematics faculty of the University of California at Berkeley. I am happy to report that this does not mark his retirement from research; he is still "covering page after page with ink," as he puts it. This means that his ideas will continue to play a seminal role in the development of general relativity, as they have in the past.

Abraham Taub was born in Chicago in 1911 and obtained his undergraduate degree from the University of Chicago at the age of twenty. He then went to Princeton for graduate study in mathematical physics. His thesis advisor was the famous cosmologist H. P. Robertson. Taub's first outstanding achievements are contained in a series of papers on spinors and the Dirac equation written between 1934, when he had just received his Ph.D., and 1950. Many of the ideas underlying twistor theory are implicit in these early papers. In particular, what is now known as local twistor theory was developed in the 1930s by Veblen, von Neumann, and Taub. He also did fundamental work on the action of the conformal group on spinors; this work is basic to modern developments in twistor theory.

Relativistic hydrodynamics has always been a subject dear to Taub's heart. In fact, many basic results on special relativistic fluid flows are due to him, and he has been a major contributor to the study of fluid flows near shocks. The importance of Taub's work in relativistic hydrodynamics is pointed out in the paper by Smarr, Taubes, and Wilson in this book. His interest in hydrodynamics led him to develop variational methods in general relativity and fluid dynamics. His research on variational methods began with his famous 1954 paper on perfect fluids and still continues with several key review articles written in recent years. The so-called Taub conserved quantities arose from Taub's work on variational methods, and these quantities are playing an essential role in the development of linearization stability theory, as discussed in the contribution by Fischer, Marsden, and Moncrief to this volume.

Taub's single most famous paper is his 1951 *Annals of Mathematics* article entitled "Empty spacetimes admitting a three parameter group of motions." This paper gives several spatially homogeneous vacuum solutions to Einstein's equations, one of which, when extended to "Taub-NUT" space, is well known to every relativist. This solution has played a key role in the development of general relativity, for in the words of C. W. Misner, it "is a counter-example to almost anything." By providing examples of pathologies that can occur in Einstein universes, Taub-NUT space has been crucial to the understanding of singularities. Taub space is the simplest known exact vacuum solution with three-sphere spatial topology and thus provides an easily understood model of possible behavior in closed universes, as the papers by Wheeler and Shepley demonstrate. More generally, Taub's 1951 paper gave the essential methods used in discusssing homogeneous aniso-trophic cosmological models. See for example the papers by Jantzen and MacCallum.

In addition to his work in classical general relativity, Taub has been active in computer science and analysis; he was instrumental in setting up the computer systems at the University of Illinois and at Berkeley. He was one of the founders of the now burgeoning field of numerical general relativity.

Throughout his career, Taub has grappled with the problem of singularities in general relativity. He has persuasively argued that the time has come to attack the problem of singularities from the point of view of distribution theory and that action integrals can be used to define space–time structures at singularities. Some of my own current work is concerned with this action integral idea of Taub's, and I feel that Taub's approach to singularities will be an important complement to the more popular approach of Hawking and Penrose.

Taub has also had an important influence on general relativity through his penetrating discussions with his colleagues. These discussions are often heated, for Taub is very definite about expressing his point of view. This definiteness has occasionally disconcerted scientists who do not know him well, and who thus are not aware of his great inner gentleness. In the words of his last graduate student, Robert Jantzen, "Professor Taub has a crusty exterior, but underneath he is just one big marshmallow!"

Some of the papers in this Festschrift were delivered at the Berkeley Symposium on Relativity and Cosmology, held in August 1978 to honor Professor Taub. I am grateful to my colleagues on the organizing committee of that Symposium, D. Eardley, E. Liang, J. Marsden, and V. Moncrief for their assistance in preparing this volume.

FRANK J. TIPLER

June 15, 1979

1

On Schwarzschild Causality—A Problem for "Lorentz Covariant" General Relativity

Roger Penrose

Mathematical Institute
University of Oxford
Oxford, England

and

Department of Mathematics
University of California
Berkeley, California

I. Introduction

There appears to be a viewpoint, prevalent among some physicists [1] (cf. [2]), that while a geometrical approach to general relativity may have merits on aesthetic grounds and may have appeal for those whose interests are, perhaps, essentially pure mathematical, a strong emphasis on curved-space geometry is nevertheless to be rejected if real physical understanding and important future progress are to be achieved in gravitation theory. Abe Taub, however, is clearly not of this way of thinking. His many important contributions to both the physical and geometrical aspects of general relativity bear strong witness to the fact that far from being an obstacle to progress, differential geometry is an efficient and essentially indispensable tool in this highly significant aspect of physical insight.

As a way of honoring Abe's retirement, I shall present here a result that lends strong additional support to this geometrical viewpoint. It is directed, particularly, against the idea that general relativity might be adequately described as though it were a Lorentz-covariant (or, more correctly, Poincaré-covariant) field theory according to which the physical metric tensor is to be

1

treated as though it were not significantly different from any other field tensor.

II. Lorentz Covariance and Causality

Now the fundamental and unique role played by the physical metric tensor g (or at least by its conformal part $\hat{g}$, which represents 9 out of its 10 algebraically independent components) is, indeed, that it defines the *physical causal relations between points*. These causal relations play a key role in any classical relativistic field theory since they determine the propagation directions for *all* relativistic fields. Furthermore, the significance of this causal structure is as great in quantum field theory as in classical field theory. Quantum causality has the implication that field operators at spacelike-separated points must necessarily commute. If we were to take the standard "Lorentz-covariant" view, then we would need to introduce a background Minkowski metric η (perhaps not canonically) with the property that any two field operators at points that are spacelike-separated with respect to the *flat* causal structure defined by η would necessarily commute. (To modify this rule would be to reject the standard Lorentz-covariant viewpoint, from which all the standard results, such as the PCT and spin-statistics theorems are derived.) This is not to say that the final causality that is physically observed need agree with that defined by η. The actual way that fields propagate in the resulting theory would have to be calculated in detail. A normal procedure for doing this would be to obtain the metric g from a power series expansion of Lorentz-covariant terms, this being an infinite summation of Feynman graphs. (Summing "tree diagrams" is to give the classical g-field.) If such a Lorentz-covariant theory is to agree with general relativity, then the finally derived field propagation has to follow the null cones of this resulting general-relativistic curved metric g instead of those of η.

For a satisfactory theory, however, one would anticipate an important consistency requirement relating η to g: *the causality defined by g should not violate the background η-causality*. To put this another way, the g-null-cones ought never to extend outside the η-null-cones (Fig. 1). Thus, timelike curves with respect to g should always remain "timelike" with respect to η, i.e.,

$$g(T, T) > 0 \Rightarrow \eta(T, T) > 0 \qquad \text{at every point,} \qquad \text{(II.1)}$$

for every tangent vector T (using the Lorentzian signature $+ - - -$). I write this condition

$$g < \eta. \qquad \text{(II.2)}$$

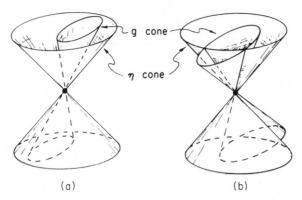

Fig. 1 Physical g-causality should not violate background η-causality. (a) Allowed: $g < \eta$ holds and (b) forbidden: $g < \eta$ fails.

If this condition were to *fail* to hold, then *physical* field propagation, which will follow the g-cones (or arbitrarily closely to them), would be superluminary with respect to η-causality. If we anticipate "physical" quantum field operators, describing the physically measurable fields, and constructible (say by infinite summation) from the original Lorentz-covariant operators, then these new field operators ought to be noncommuting for g-null- (or perhaps g-timelike-) separated points. But if $g < \eta$ fails, then these operators would have also to be noncommuting at certain pairs of η-spacelike-separated points (namely those along g-cones that extend outside η-cones). But this is not possible since noncommuting operators cannot be built out of the original commuting ones at spacelike η-separation.

Of course, in a proper quantized theory of general relativity, in which g also becomes a quantized field, the "g-cones" would never be perfectly well defined. However, if such a theory were to bear any resemblance to standard general relativity in the *classical limit*, then there should be a resulting "approximate" classical g-metric that would be anticipated to satisfy $g < \eta$, for the reasons outlined above. In the absence of a good theory of quantized gravity, there necessarily remains a certain inconclusiveness in this argument. Nevertheless, a question of some considerable interest for its *own* sake is whether or not, for a physically reasonable metric g on a space–time manifold M, a Minkowski metric η also exists on M with $g < \eta$. And if not, then this fact would also seem to provide some pertinent evidence against the fruitfulness of a Lorentz-covariant approach.

It is clear that for certain "strong gravitational fields" we should have difficulty in arranging $g < \eta$. For example, the *topology* of the space–time might itself differ from that of Minkowski space. Also there are space–times with causality violations (such as a maximally extended Kerr solution [3]) and certain examples (such as plane waves [4]), having no Cauchy hyper-surfaces, and for which, for rather blatant reasons, $g < \eta$ cannot be arranged. But all these examples could be reasonably argued to be "nonphysical" and therefore not to constitute any significant case against the Lorentz-covariant viewpoint. However, I shall show here that the situation is much more serious than this. In even the simplest of physically reasonable nontrivial space–time geometries, namely, the Schwarzschild solution (with any suitable interior) $g < \eta$ indeed *cannot* be *satisfactorily* arranged.

III. Schwarzschild Geometry

At first, it would seem that the standard Schwarzschild coordinate system (t, r, θ, ϕ), for which the g-metric takes the form

$$ds^2 = \left(1 - \frac{2m}{r}\right)dt^2 - \left(1 - \frac{2m}{r}\right)^{-1} dr^2 - r^2(d\theta^2 + \sin^2\theta\, d\phi^2), \quad \text{(III.1)}$$

provides a counter example to above contention, since all the null cones of this metric do in fact lie inside those of the corresponding Minkowski metric (in spherical polar coordinates):

$$d\sigma^2 = dt^2 - dr^2 - r^2(d\theta^2 + \sin^2\theta\, d\phi^2) \quad \text{(III.2)}$$

provided that we restrict to the range

$$r \in [r_0, \infty) \quad \text{(III.3)}$$

for some $r_0 > 2m$. In order to cover the remaining range, we need an "interior" solution, such as the constant density perfect fluid solution originally proposed by Schwarzschild [1]

$$ds^2 = \left[\frac{3}{2}\left(1 - \frac{2m}{r_0}\right)^{1/2} - \frac{1}{2}\left(1 - \frac{2mr^2}{r_0^3}\right)^{1/2}\right]^2 dt^2$$

$$- \left[1 - \frac{2mr^2}{r_0^3}\right]^{-1} dr^2 - r^2(d\theta^2 + \sin^2\theta\, d\phi^2), \quad \text{(III.4)}$$

where

$$r \in [0, r_0], \quad r_0 > 3m. \quad \text{(III.5)}$$

(I am not concerned with the question of black holes here, which would lead to some added complications.) These two solutions (III.1) and (III.4) match together to give a manifold with a C^0-metric g for which $g < \eta$, where the flat metric is given by (II.2). [It would not be hard, in fact, to modify the interior metric (III.4) to obtain a C^∞-metric g with this property, especially if we are not concerned with any particular equation of state in the interior region.] We might think, therefore, that the effect of the Schwarzschild gravitational field is simply to "slow down" the velocity of light and so to provide $g < \eta$.

There is, however, an important sense in which this setup is quite unsatisfactory. This relates to behavior at infinity. One of the reasons for adopting a Lorentz-covariant viewpoint would, after all, be to discuss scattering theory. We might be concerned with incoming and outgoing asymptotically plane waves (e.g., gravitational perturbations or test Maxwell or massless scalar fields). Alternatively, taking the geometrical optics limit, we could be concerned with incoming and outgoing null geodesics. But, with regard to these null geodesics and null surfaces, the g-metric given by (III.1) actually differs very greatly from the η-metric given by (III.2) at large distances from the matter source (despite the fact that the metrics appear, naïvely, to go into one another in the limit $r \to \infty$). Consider, for example, the radial outgoing g-null geodesics given by

$$u = \text{const}, \qquad \theta = \text{const}, \qquad \phi = \text{const},$$

where u is g-retarded time:

$$u = t - r - 2m \log(r - 2m). \tag{III.6}$$

It is clear from the form of (III.6) that as $r \to \infty$, the value of $t - r$ is unbounded above along the geodesic. But $t - r$ is the standard η-retarded time, so we see that outgoing g-null geodesics reach indefinitely far into the η-retarded future, and do not correspond at all to outgoing η-null geodesics. Correspondingly, the incoming g-null geodesics are generators of the incoming g-null cones along which the g-advanced time

$$v = t + r + 2m \log(r - 2m) \tag{III.7}$$

is constant. Again there is no correspondence with the incoming η-null geodesics since the η-advanced time $t + r$ becomes unboundedly large and negative into the past along a g-null geodesic.

This is clearly unsatisfactory if this η-metric is to be used in a Lorentz-covariant approach for studying scattering theory in a Schwarzschild space–time. On the other hand, we might envisage using an η-metric which is related to the Schwarzschild g-metric in a *different* way from the one just considered. To put things another way, we might choose a different coordinate system for the space–time, whose naturally associated flat metric was not the same as the η just considered. For example, we could use

$$r' = \tfrac{1}{2}(v - u) = r + 2m \log(r - 2m) \tag{III.8}$$

in place of r and now adopt the flat η'-metric given by

$$d\sigma'^2 = dt^2 - dr'^2 - r'^2(d\theta^2 + \sin^2 \theta \, d\phi^2). \tag{III.9}$$

In this particular case, since $u = t - r'$ and $v = t + r'$ are retarded and advanced times, respectively, for *both* metrics, the difficulty that was just mentioned concerning the difference in asymptotic structure between g and η does *not* now arise. However, it turns out that the condition $g < \eta$ is now what fails!

The purpose of this article is to show that there is, indeed, an essential incompatibility between the causal structure in Schwarzschild space–time (of positive mass m, and with *any* interior whatever) and that of Minkowski space. This shows up either asymptotically or in a violation of the local condition $g < \eta$. The most convenient way to handle the asymptotic conditions is by use of *conformal infinity* [3,5,6]. Both the Schwarzschild and Minkowski space–times have, in fact, well-behaved conformal infinities $\mathscr{I}^{\pm}$. Each point p of $\mathscr{I}^{+}$ can be thought of as describing an *outgoing (null) asymptotically plane wavefront*. (This is [3,7] the boundary $\partial I^{-}(p)$ of the TIP $I^{-}(p)$, cf. (IV. 2), representing p—an "event horizon" [6].) Likewise each point q of $\mathscr{I}^{-}$ describes an *incoming (null) asymptotically plane wavefront* (the boundary $\partial I^{+}(q)$ of the TIF $I^{+}(q)$ representing q—a "creation horizon" [3], sometimes referred to as a "particle horizon" [6]). Thus, $\mathscr{I}^{+}$ and $\mathscr{I}^{-}$ have very direct and natural interpretations in the context of scattering theory.

The space–time manifold with both conformal infinities $\mathscr{I}^{\pm}$ adjoined will be denoted $\overline{M}$. The conformal metric $\hat{g}$ is well-defined (C^{0} will do) on the whole of $\overline{M}$, including its boundary $\mathscr{I}^{-} \cup \mathscr{I}^{+}$. The required condition that the asymptotic η-causal structure agree with the asymptotic g-causal structure (i.e., that the scattering theory be compatible for both metrics) is that both $\hat{g}$ and $\hat{\eta}$ be well defined on the *same* manifold-with-boundary $\overline{M}$. We have seen that this does *not* in fact hold for the $\hat{g}$ and $\hat{\eta}$ conformal metrics related to each other as is entailed by (III.1) and (III.2), but that it *does* hold for $\hat{g}$ and $\hat{\eta}$, related by (III.1) and (III.9).

IV. The Theorem

Let us now consider the more general question of whether or not a flat conformal $\hat{\eta}$-metric [not necessarily related to (III.1) by (III.2) or by (III.9)] exists at all on $\overline{M}$ and for which the required condition

$$\hat{g} < \hat{\eta} \tag{IV.1}$$

holds [this being the same as (II.2), but written now in terms of the conformal metrics, to emphasize that it applies to the *whole* of $\overline{M}$]. Here $\overline{M}$, with its given conformal metric $\hat{g}$, refers to the standard positive-mass exterior Schwarzschild solution [given by (III.1) in the range (III.3)] with any suitable interior [such as (III.4) in the range (III.5)—but the choice of interior turns out to be irrelevant] and with standard conformal boundary $\mathscr{I}^+ \cup \mathscr{I}^-$ [with the points of $\mathscr{I}^+$ given by (u, θ, ϕ) at $r = \infty$ and those of $\mathscr{I}^-$ given by (v, θ, ϕ) at $r = \infty$]. The topology of $\overline{M}$ is built up from the (Schwarzschild) space–time:

$$M = \text{int } \overline{M} \cong \mathbb{R}^4$$

and its conformal boundary $\partial \overline{M} = \mathscr{I}^+ \cup \mathscr{I}^-$:

$$\mathscr{I}^+ \cong \mathscr{I}^- \cong S^2 \times \mathbb{R}.$$

With the interior (III.4) in the range (III.5), M is actually an asymptotically simple space–time [3,5,6]. The standard notation $a \ll b$, for points $a, b \in \overline{M}$ is adopted [6,8] for the assertion "there exists a future-directed timelike curve in $\overline{M}$ from a to b." Also

$$I^+(a) = \{x \in \overline{M} \,|\, a \ll x\}, \qquad I^-(a) = \{x \in \overline{M} \,|\, x \ll a\}. \qquad \text{(IV.2)}$$

I shall prove:

(IV.3)   *Theorem* $\mathscr{I}^+ \subset I^+(a)$ *for each* $a \in \mathscr{I}^-$; *equivalently*, $\mathscr{I}^- \subset I^-(a)$ *for each* $b \in \mathscr{I}^+$.

An equivalent statement but which does not refer to $\mathscr{I}^\pm$ is:

(IV.4)   *Theorem*   *If* λ *and* μ *are endless timelike curves in* M, *then there exist points* $p \in \lambda$, $q \in \mu$ *with* $p \ll q$.

From these results can be derived, as a simple corollary, the required property

(IV.5)   *Theorem*   *There is no Minkowskian conformal metric* $\hat{\eta}$ *on* $\overline{M}$ *with* $\hat{g} < \hat{\eta}$.

Proofs   Let us first establish the equivalence of (IV.3) with (IV.4). Note that another way of stating (IV.3) is

$$a \ll b, \text{ for every pair } a \in \mathscr{I}^-, b \in \mathscr{I}^+. \qquad \text{(IV.6)}$$

Now, using the notion of causal boundary [3,7], we can interpret the point $a \in \mathscr{I}^-$ as an equivalence class of endless timelike curves having the same future in M (their points constituting the TIF in M representing a). Likewise, $b \in \mathscr{I}^+$ can be interpreted as an equivalence class of endless timelike curves with the same past in M (generating the TIP representing b). Let the endless timelike curves λ and μ represent a and b, respectively, so we can think of λ as acquiring a past endpoint a on $\mathscr{I}^-$ and μ as acquiring a future endpoint b on $\mathscr{I}^+$. The assertion $a \ll b$ amounts to saying that an endless timelike curve v exists whose future agrees with that of λ and its past with that of μ. Now it is clear that *if* the statement of (IV.4) holds, then v does indeed exist, namely, consisting of the past-endless portion of λ up until the point p, the asserted timelike curve from p to q, and the future-endless portion of μ from q onward. (This jointed curve can be smoothed, if desired.) Conversely, suppose that λ and μ are as above and that ν has the same future as λ and the same past as μ. Choose a point w on ν. Now w must lie in the future of λ (since it lies in the future of ν), so a point p exists on λ with $p \ll w$. Similarly, q exists on μ with $w \ll q$. Hence $p \ll q$ holds, as required in (IV.4). There remains the possibility that one or both of the curves λ, μ in (IV.4) do not reach $\mathscr{I}^{\pm}$ at all, at their appropriate ideal endpoints but reach timelike infinity $i^{\pm}$ instead [3,5–7]. In these cases (IV.4) is again satisfied, but in a more trivial way.

Next, let us see why (IV.5) is a consequence of (IV.3) or (IV.4). Suppose a Minkowskian $\hat{\eta}$ exists on $\overline{M}$ with $\hat{g} < \hat{\eta}$. Now it is clear that (IV.4) would be false if "timelike" and " $\ll$ " referred to the flat η metric. For example, in the usual T, X, Y, Z coordinates, with $d\sigma^2 = dT^2 - dX^2 - dY^2 - dZ^2$, the two branches of the η-timelike hyperbola $T^2 - X^2 + 1 = 0 = Y = Z$ are everywhere η-spacelike separated from one another. Suppose we let a be the past endpoint that one branch acquires on $\mathscr{I}^-$ and b the future endpoint that the other branch acquires on $\mathscr{I}^+$. By (IV.6), there is a g-timelike curve in $\overline{M}$ from a to b. But if $\hat{g} < \hat{\eta}$, this curve must also be η-timelike. But this is impossible as we have just seen. [The equivalence between (IV.4) and (IV.6) clearly holds equally for the $\hat{\eta}$-causality as for the $\hat{g}$-causality.] This contradiction establishes (IV.5) as a consequence of (IV.3) or (IV.4).

Finally, we must establish (IV.3). Let c be a point in M (with $r > r_0 > 3m$) and let γ be a null geodesic through c which is transverse to the source at c. Without loss of generality we can arrange c to have coordinates $(0, R, \pi/2, \pi/2)$ in the standard (t, r, θ, ϕ) coordinate system of (III.1) and, with the dot denoting $d/d\tau$, where τ is an affine parameter on γ,

$$\dot{r} = \dot{\theta} = 0 \qquad \text{at } t = 0. \tag{IV.7}$$

The standard geodesic equations for γ yield

$$\theta = \pi/2, \qquad r^2\dot{\phi} = A = \text{const}, \tag{IV.8}$$

and

$$[1 - (2m/r)]\dot{t} = B = \text{const}. \tag{IV.9}$$

We can normalize the affine parameter τ by choosing

$$B = 1 \qquad \text{and} \qquad \tau = 0 \qquad \text{at } t = 0. \tag{IV.10}$$

Then, by (IV.7)–(IV.10) (and taking the positive sign for $\dot{\phi}$),

$$A = R[1 - (2m/R)]^{-1/2}$$

and [using (III.6): $u = t - r - 2m \log(r - 2m)$], we derive

$$u = -R - 2m \log(R - 2m) + \int_R^r f(\rho)d\rho$$

where

$$f(\rho) = \left(1 - \frac{2m}{\rho}\right)^{-1} \left\{ \left[1 - \frac{R^2}{\rho^2} \left(\frac{1 - 2m/\rho}{1 - 2m/R} \right) \right]^{-1/2} - 1 \right\}.$$

By inspection of (IV.13) we see that

$$\int_R^\infty f(\rho)d\rho \tag{IV.11}$$

converges—as it must, since γ reaches a point of $\mathscr{I}^+$ with finite u-value. We wish to examine how this u-value behaves as $R \to \infty$ and to show, in fact, that it tends to $-\infty$:

$$\lim_{R \to \infty} \left[-R - 2m \log(R - 2m) + \int_R^\infty f(\rho)d\rho \right] = -\infty. \tag{IV.12}$$

Now, to see this, we note

$$f(\rho) < \left(1 - \frac{2m}{R}\right)^{-1} \left\{ \left[1 - \frac{R^2}{\rho^2(1 - 2m/R)} \right]^{-1/2} - 1 \right\} = g(\rho)$$

if

$$\rho \in (P, \infty) \qquad \text{where} \qquad P = R \left(1 - \frac{2m}{R} \right)^{-1/2} = R + O(1),$$

whence

$$\int_P^\infty f(\rho)d\rho < \int_P^\infty g(\rho)d\rho = R \left(1 - \frac{2m}{R} \right)^{-3/2} = R + O(1) \tag{IV.13}$$

by explicit integration. Furthermore we can estimate $\int_R^P f(\rho)d\rho$ using

$$f(\rho) = \left(1 - \frac{2m}{\rho}\right)^{-1}$$
$$\times \left(\left\{ \frac{(\rho - R)[\rho R(\rho + R) - 2m(\rho^2 + \rho R + R^2)]}{\rho^3(R - 2m)} \right\}^{-1/2} - 1 \right)$$
$$< \left(1 - \frac{2m}{R}\right)^{-1} (\rho - R)^{-1/2} P^{3/2} (R - 2m)^{1/2} (2R^3 - 6mP^2)^{-1/2},$$

provided that

$$\rho \in (R, P) \qquad \text{and} \qquad R > 5m$$

whence, using

$$\int_R^P (\rho - R)^{-1/2}\, d\rho = 2(P - R)^{1/2} = O(1),$$

we derive the fact that

$$\int_R^P f(\rho) d\rho = O(1) \tag{IV.14}$$

as $R \to \infty$. Combining (IV.14) with (IV.13), we obtain the result that (IV.11) is $R + O(1)$, so that substituting into the left-hand side of (IV.12), we derive the required result (IV.12), because of the presence of the logarithmic term.

This shows that whatever value of u is chosen, R can be made large enough that the null geodesic γ meets $\mathscr{I}^+$ at a u-value that is *less* than that chosen value. Moreover, because of the light-bending effect, γ will encounter an outgoing radial null geodesic β whose equation has the form

$$u = u_0 = \text{const}, \qquad \theta = \pi/2, \qquad \phi = \phi_0 = \text{const} \tag{IV.15}$$

for any value of ϕ_0 in the range

$$\phi_0 \in [\pi/2, \pi] \tag{IV.16}$$

(*including* the value $\phi_0 = \pi$), before reaching $\mathscr{I}^+$. Since u is an increasing function along γ, the value u_0 must be even less than the u-value attained at $\mathscr{I}^+$. We can likewise repeat the entire preceding argument in time-reversed form and attain the result that whatever value of v [given by (III.7)] is chosen, R can be made large enough that γ encounters, into the past, an incoming null geodesic α with equation

$$v = v_1 = \text{const}, \qquad \theta = \pi/2, \qquad \phi = \phi_1 = \text{const} \tag{IV.17}$$

for which v_1 is larger than the chosen v-value, and where ϕ_1 can take any value in the range

$$\phi_1 \in [0, \pi/2].$$

A jointed null geodesic, made up from pieces of α, γ, and β therefore connects the point $a \in \mathscr{I}^-$, with (v, θ, ϕ)-coordinates $(v_1, \pi/2, \phi_1)$, to the point $b \in \mathscr{I}^+$, with (u, θ, ϕ)-coordinates $(u_0, \pi/2, \phi_0)$. Thus [8]

$$a \ll b$$

as is required for (IV.6). (The jointed null geodesic can be smoothed, if desired, to yield to smooth timelike curve from a to b.) By a suitable rotation of the (θ, ϕ)-coordinate system, we can arrange for a to lie on any generator of $\mathscr{I}^-$ and b on any generator of $\mathscr{I}^+$. (The crucial case, in fact, is when the generators are opposite: $\phi_0 = \pi$, $\phi_1 = 0$.) And by allowing v_1 to be as large and positive as desired, and u_0 to be as large and negative as desired, we can cover all possibilities, thus establishing (IV.3). ∎

V. Concluding Remarks

It should be clear from the preceding construction that the results of this paper are in no way specific to the Schwarzschild solution. One is concerned only with the nature of the space–time at large distances from the positive-mass source—evidently essentially with causal properties in the neighborhood of spacelike infinity i^0 [3,5]. Corresponding results are to be anticipated for *any* appropriately asymptotically flat space–time with positive mass.

One is tempted to use the fact that, whenever the null convergence condition holds [3,9] (a consequence of the weak energy condition and Einstein's equations) together with the genericity condition [3,9], every complete null geodesic in the space–time contains pairs of conjugate points [3,9]. This has the implication that for any point $a \in \mathscr{I}^-$ (assuming asymptotic simplicity), no generator of $\partial I^+(a)$ in the space–time extends all the way to $\mathscr{I}^+$. This imposes severe difficulties for the geometry of $\partial I^+(a) \cap \mathscr{I}^+$, if this set is to be nonvacuous, and appears to lead to a more general argument from which (IV.3) can be derived under much wider circumstances. The question seems also to be related to the positive energy conjecture [10,11] and to the details of the structure [12] of i^0. These matters will not be discussed here, as the result obtained in Section IV is adequate for the present purposes.

Acknowledgment

The author is grateful to the Miller Foundation and the University of California at Berkeley for their support and hospitality.

References

[1] Weinberg, S., "Gravitation and Cosmology." Wiley, New York, 1972.
[2] Duff, M. J., *in* "Quantum Gravity." C. J. Isham, R. Penrose, and D. W. Sciama, eds., p. 78–135. Oxford Univ. Press, London and New York, 1975.
[3] Hawking, S. W., and Ellis, G. F. R., "The Large Scale Structure of Spacetime." Cambridge Univ. Press, London and New York, 1973.
[4] Penrose, R., *Rev. Mod. Phys.* **37**, 215 (1965).
[5] Penrose, R., *Proc. R. Soc. London, Ser. A* **284**, 159 (1965).
[6] Penrose, R., *in* "Battelle Rencontres" (C. M. DeWitt and J. A. Wheeler, eds.), p. 171–189. Benjamin, New York, 1968.

[7]   Geroch, R. P., Kronheimer, E. H., and Penrose, R., *Proc. R. Soc. London, Ser. A* **327**, 545 (1972).

[8]   Penrose, R., "Techniques of Differential Topology in Relatively." *SIAM*, Philadelphia, Pennsylvania, 1972.

[9]   Hawking, S. W., and Penrose, R., *Proc. R. Soc. London, Ser. A* **314**, 529 (1970).

[10] Geroch, R. P., *Ann. N.Y. Acad. Sci.* **224**, 125 (1973).

[11] Schoen, R., and Yau, S.-T., *Commun. Math. Phys.* **65**, 45 (1979).

[12] Ashtekar, A., and Hansen, R. O., *J. Math. Phys.* **19**, 1542 (1978).

2

Comments on the Topology of Nonsingular Stellar Models†

Lee Lindblom

Department of Physics
University of California
Santa Barbara, California

and

Dieter R. Brill

Department of Physics and Astronomy
University of Maryland
College Park, Maryland

Abstract

We consider the question, what topologies of space–time are permitted by Einstein's equations as models of nonsingular stars. A number of results from the literature on singularity theorems are drawn together here to give a clear answer to this question. If a star evolves from nearly Newtonian initial conditions, of low densities and small space–time curvatures, to a nonsingular final state, then the topology of the space–time representing the star must be R^4. We also discuss some topological constraints on models that are not nearly Newtonian in the past, but may have evolved directly from the initial cosmic singularity.

I. Introduction

Professor Taub's contributions to the study of relativistic fluid mechanics form an important foundation on which any fundamental study of relativistic stellar structure must be based. We feel it is appropriate,

† Supported in part by the National Science Foundation.

therefore, to consider here another basic aspect of the study of relativistic stars: the topology of the space–time manifold representing a stellar model.

The endproducts of stellar evolution can be divided into two classes: those that contain space–time singularities and those that do not. If the cosmic censorship hypothesis [1] is true, the singular endproducts of stellar evolution are black holes. The final equilibrium black hole solutions to Einstein's equations are now completely understood due to the theorems of Israel [2], Carter [3], Hawking [4], and Robinson [5]. Therefore, it is of interest to inquire about the properties of the solutions to Einstein's equations that represent the nonsingular endpoints of stellar evolution. While some properties of these solutions are known (see Lindblom [6], for a review), there is much that is not known with certainty. We shall consider here one of the most fundamental aspects of this inquiry: What space–time topologies are possible for nonsingular solutions to Einstein's equations that represent stellar models?

Geometric theories of gravitation, such as general relativity, allow (in principle) a very large and diverse set of possible space–time topologies. One can imagine, for example, configurations of matter that have nontrivial spatial topologies such as those depicted in Fig. 1. One would like to know whether configurations of this sort are possible models for nonsingular stars. Is it possible that in regions of high space–time curvature, such as the inside of a neutron star, the topology of space–time there might be a "wormhole" as in Fig. 1a? What effect would such nontrivial topology have on the theorems that give an upper limit to the possible mass of a neutron star? Another nontrivial possibility is illustrated in Fig. 1b, and unlimited other possibilities exist. We argue here that configurations such as those in Fig. 1

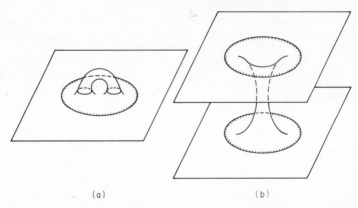

(a)   (b)

Fig. 1 Possible spatial topologies for stellar models. (a) represents a stellar model containing a nonsimply connected "wormhole" within it. The shaded line represents the surface of the star. (b) represents a model having two asymptotic regions connected by a "wormhole."

are not possible general relativistic stellar models having no singularities. We consider two separate cases. The first case corresponds to stars that originated from nearly Newtonian configurations that had low densities and small space–time curvatures. This case corresponds to the usual picture of the early stages of stellar evolution. For models of this type, we assume that past regions of the space–time have many of the asymptotic features of flat Minkowski space. These assumptions are justified for the description of normal stellar evolution because the effects of the overall curvature of our universe are quite small on distance–time scales that are still very large compared to the size–age of any star. However, it is possible that some primordial condensations of matter could have occurred very early in the history of our Universe. Perhaps these primordial stars now form the cores of quasars or even stranger as yet undiscovered objects. The possible topologies of these primordial stars represent the second case we wish to consider here.

II. The Topology of Initially Newtonian Stars

The first theorem that placed restrictions on the possible topology of stars, which began under nearly Newtonian circumstances, was given by Geroch [7]:

Theorem 1 If the space–time (M, g) is asymptotically simple and empty, then M is homeomorphic to R^4.

This theorem is extremely general and excludes nonstandard topologies provided the star never becomes extremely compact. But for compact objects, the assumption of asymptotic simplicity is stronger than is justifiable. Asymptotic simplicity requires that every null geodesic in the space–time begin and end at null infinity. This assumption would be violated by any spherical star of radius less than $3M$, because the exterior Schwarzschild geometry has circular null geodesics at $r = 3M$. Since it is possible to construct nonsingular models of radius smaller than $3M$ (see, e.g., Bondi [8]), we would hope to find results that would place restrictions on the topology of these highly compact objects as well.

Another theorem of Geroch [9] gives some topological information about a wider class of space–times than the asymptotically simple ones considered in Theorem 1.

Theorem 2 If a space–time (M, g) admits a Cauchy surface S, then M must be homeomorphic to $R \times S$.

In such a space–time, then, a star of initially trivial topology cannot change its topology even if it later becomes highly compact. However, again the assumptions on the space–time are stronger than are physically justifiable. The existence of a Cauchy surface in a space–time not only rules out the possibility of "naked singularities" (an acceptable omission for our present purposes) but also the possibility of other causal peculiarities such as closed timelike lines. One should not exclude, a priori, the possibility of bizarre causal behavior in a region of space–time having large curvatures and possible nontrivial topological structure.

In the case of a star that begins as a highly diffuse cloud of gas and dust (the usual picture), and which evolves to a nonsingular endpoint, we can show that bizarre causal behavior does not occur, i.e., that a Cauchy surface does exist. Early in the evolution of a star of this sort the density of matter is very low, and the geometry of space–time is nearly that of flat Minkowski space. For a space–time describing a star that begins in this way, it seems reasonable to assume that the space–time begins and remains asymptotically flat throughout the evolution of the star. Also, it seems reasonable to assume that any bizarre causal behavior occurs only during the late stages of stellar evolution when the densities and curvatures are high. Thus we assume the space–time has at least a partial Cauchy surface at sufficiently early times. The following theorem shows that a nonsingular stellar model that begins in this way cannot develop bizarre causal behavior as it evolves:

Theorem 3   *If a space–time (M, g) satisfies the following conditions*:

(a) (M, g) *is geodesically complete.*

(b)   *The weak energy condition and Einstein's equations hold on (M, g).*

(c)   *The generic condition holds on (M, g).*

(d) (M, g) *is weakly asymptotically simple and empty.*

(e) (M, g) *is partially asymptotically predictable from a partial Cauchy surface S.*

Then S is a Cauchy surface for (M, g).

The proof of this result is virtually identical to the proof of a closely related result by Tipler [10], and the interested reader is referred to the proof of his Theorem 1. (Another related result is given by Tipler [11], Theorem 6.) Assumptions (a)–(e) in our result are the technical restrictions that reflect the physical situation described qualitatively above. The precise mathematical meanings of these terms can be found in Hawking and Ellis [12] and Tipler [10]. For the reader unfamiliar with the language of "global techniques," we describe qualitatively the meaning of each assumption:

(a)   Geodesic completeness is the condition that the star have no singularities.

(b)   The weak energy condition requires that the density of matter must never be negative.

(c)   The generic condition is a technical assumption that requires that every geodesic observer feel some tidal force sometime during its history. It is expected to be satisfied by physically realistic space–times.

(d)   Weak asymptotic simplicity is the requirement that the space–time behave asymptotically (near null finity) like flat Minkowski space.

(e)   The existence of a partial Cauchy surface, from which the space–time is partially asymptotically predictable, excludes any causal anomalies from early times in the history of the stars and from the asymptotic region of the space–time near spacelike infinity.

We can conclude from Theorems 2 and 3 that the space–time of a non-singular star, which evolves from an initial state of low density, must have the topology $R \times S$. Moreover, we can determine what topology S must have. Consider one of the partial Cauchy surfaces S, which occurs very early in the evolution of the star when the densities are very low. It is reasonable to assume that the past directed null geodesics leaving this surface all reach past null infinity, since the space–time to the past of S is nearly the same as Minkowski space. Such a partial Cauchy surface is said to have an asymptotically simple past (see Hawking and Ellis [12], p. 316). By the same arguments used (by Geroch [7]) to prove Theorem 1, it follows that S must be homeomorphic to R^3, and consequently M must be homeomorphic to R^4. Thus we have:

Theorem 4   If a space–time (M, g), having a partial Cauchy surface S with an asymptotically simple past, satisfies the assumptions of Theorem 3, then M is homeomorphic to R^4.

Consequently, for any nonsingular stellar model in Einstein's theory that evolves from low density, nearly Newtonian, initial conditions must have the trivial topology R^4.

III.   The Topology of Primordial Stars

We next consider the more difficult case of primordial stars, those (if any) that were formed immediately after the initial singularity of our universe, rather than from nearly Newtonian initial conditions. What assumptions can we make about the asymptotic features of the space–times describing these objects? The primordial objects, which we have in mind, are those that look from a distance like the exteriors of ordinary stars (clusters, or galaxies) but whose interiors might be bizarre, because they arose from the early stages of our universe rather than from nearly Newtonian conditions. Thus it is reasonable to assume that the asymptotic properties of the manifold are weakly asymptotically simple and empty, *except* possibly for some open neighborhood of past timelike infinity i_-. It may be, for example, that past null infinity $\mathscr{I}^-$ in these models is not complete in the past. Although there is no strong motivation for placing restrictions on possible bizarre causal behavior that may have been associated with these objects from the earliest times, it is reasonable to limit the spatial extent of any such peculiarities. Since we do not observe causal anomalies in the laboratory, any such anomalies associated with primordial stars should be confined spatially to regions of space–time near the interiors of these objects.

While no restrictions on the topologies of primordial stars (which satisfy only the above assumptions) have yet been found, some progress has been made under more restrictive conditions.

Theorem 5   If a space–time (M, g) satisfies the assumptions of Theorem 3 and the chronology condition, and admits an asymptotically regular partial Cauchy surface S, then M is homeomorphic to R × S, S is simply connected, and the boundary of S is a single two-sphere.

This theorem is proved by combining the results of our Theorem 3 with the work of Gannon [13,14]. (See also the work of Lee [15,16].) We have introduced two new assumptions in this theorem. The chronology condition assumes the nonexistence of closed timelike lines. The condition of asymptotic regularity of the surface S is a condition that requires that S be "asymptotically flat" in an appropriate sense (see Gannon [13,14]). The use of the asymptotic regularity of S and weak asymptotic simplicity in the proof of this theorem involve only the properties of the space–time near spacelike infinity i_0 and future null infinity $\mathscr{I}^+$. Consequently these assumptions are acceptable for our purposes.

Theorem 5 shows that models of primordial stars, which are nonsingular and devoid of causal anomalies, must be simply connected (thus ruling out examples like Fig. 1a) and may have only one asymptotic region (thus ruling out examples like Fig. 1b). If the Poincaré conjecture is true, these restrictions on the topology of S are sufficient to show that S is homeomorphic to R^3. Even without the Poincaré conjecture, however, the result completely determines the structure of the topological groups of S. It follows, for example, that all of the homotopy groups are trivial: $\pi_k(S) = 1$, $k \geq 0$ (see Hempel [17]). Furthermore, one can conclude that the integral singular homology groups of S must be trivial: $H_k(S) = 0$, $k \geq 1$ (use the Hurewicz isomorphism theorem, see Spanier [18], p. 398). The remaining group $H_0(S)$ is isomorphic to the group of integers under addition, $H_0(S) = Z$. The integral singular cohomology groups can then be computed using Lefschetz duality and the singular cohomology (or homology) exact sequence (see

Hu [19]). The resulting groups are $H^0(S) = Z$ and $H^k(S) = 0$, $k \neq 0$. Thus S has the same homotopy, homology, and cohomology structure as R^3.

The weakness of Theorem 5, from the viewpoint of the study of primordial stars, is the overly restrictive causal assumptions: the chronology condition and the existence of a partial Cauchy surface. The theorem does show, however, that any nontrivial topology in these models must be accompanied by bizarre causal behavior as well.

Acknowledgment

We would like to thank James Hartle for helpful comments.

References

[1] Penrose, R., *Nuovo. Cimento, Suppl.* **1**, 252 (1969).
[2] Israel, W., *Phys. Rev.* **164**, 1776 (1967).
[3] Carter, B., *Phys. Rev. Lett.* **26**, 331 (1971).
[4] Hawking, S. W., *Commun. Math. Phys.* **25**, 152 (1972).
[5] Robinson, D. C., *Phys. Rev. Lett.* **34**, 905 (1975).
[6] Lindblom, L. A., Ph.D. Thesis, Univ. of Maryland, College Park, 1978.
[7] Geroch, R., *in* "General Relativity and Cosmology" (R. K. Sachs, ed.), pp. 96–99. Academic Press, New York, 1971.
[8] Bondi, H., *Proc. R. Soc. London, Ser. A* **282**, 303 (1964).
[9] Geroch, R., *J. Math. Phys.* **11**, 437 (1970).
[10] Tipler, F. J., *Phys. Rev. Lett.* **37**, 879 (1976).
[11] Tipler, F. J., *Ann. Phys. N.Y.* **108**, 1 (1977).
[12] Hawking, S. W., and Ellis, G., "The Large Scale Structure of Spacetime." Cambridge Univ. Press, London and New York, 1973.
[13] Gannon, D., *J. Math. Phys.* **16**, 2364 (1975).
[14] Gannon, D., *Gen. Relativ. Grav.* **7**, 219 (1976).
[15] Lee, C. W., *Commun. Math. Phys.* **51**, 157 (1976).
[16] Lee, C. W., *Proc. R. Soc. London, Ser. A* **364**, 295 (1979).
[17] Hempel, J., "3-Manifolds," Annals of Mathematical Studies, No. 86, p. 26. Ann. Math., Princeton, New Jersey, 1976.
[18] Spanier, E. H., "Algebraic Topology." McGraw-Hill, New York, 1966.
[19] Hu, S. T., "Cohomology Theory." Markham Publ. Co., Chicago, Illinois, 1968.

3

General Relativity and the Eternal Return[†]

Frank J. Tipler

Center for Theoretical Physics
University of Texas
Austin, Texas

"The thing that hath been, it is that which shall be; and that which is done is that which shall be done: and there is no new thing under the sun."

Ecclesiastes 1:9

"God forbid that we should believe this. For Christ died once for our sins, and rising again, dies no more."

St. Augustine ("City of God, XII," Chap. 13)

Abstract

An arbitrarily close return to a previous initial state of the Universe, such as is predicted by the Poincaré recurrence theorem, cannot occur in a closed universe governed by general relativity. The significance of this result for cosmology and thermodynamics is pointed out.

I. Introduction

The idea that history repeats itself—that every state of the Universe has occurred before and will occur again *ad infinitum*—is a concept that originated thousands of years ago and in the form of the Poincaré

† Work supported by the National Science Foundation under grant number PHY-77-15191.

cycle and the "cyclic" universe, still plays an important role in fundamental physics. This notion of an "external return" will be studied in this paper in the context of general relativity, and it will be shown that in a closed universe, no arbitrarily close (or exact) return to a previous state of the universe is possible, provided certain very general restrictions on the matter tensor and global causal structure hold. This result is in stark contrast to the situation in classical mechanics: Poincaré [1,2] showed that for almost all initial states, any mechanical system with a finite number of degrees of freedom, finite total and kinetic energy, and constrained to move within a finite box must necessarily return arbitrarily closely and infinitely often to almost every previous state of the system. The reason for "no return" in general relativity and "eternal return" in classical mechanics is that in general relativity, singularities intervene to prevent recurrence. General relativistic closed universes are thought to begin and end in singularities of infinite curvature, and these singularities force time in general relativity to be linear rather than cyclic.

The notions of recurrence versus no-return constitute what Holton [3,4] calls a thema–antithema pair. Briefly, these are pairs of opposite fundamental concepts that form the basic framework of all scientific models. Other examples are absolute versus relative, the plenum versus the void, atomism versus the continuum, and determinism versus indeterminism. Holton argues that although scientific theories change, the fundamental thema (whose number he [4] estimates to be less than 100) remain in different guises and different proportions in succeeding theories. This is certainly true of the recurrence/no-return couple. A brief history of the eternal return concept in philosophy and science will be given in Section II, with emphasis on the versions of this idea that have strong parallels in modern science. Section III will give a precise statement of the general relativistic no-return theorem, together with a brief explanation of the meaning of the terms used in the theorem. The proof of the theorem will be given in Section IV.

As pointed out by numerous authors [2,5–7] Poincaré recurrence is a major stumbling block to the definition of entropy, a function of the state of the system that never decreases, in terms of the fundamental microscopic variables of the system. Hitherto the increase of an "entropy" defined by such variables has been obtained (not very successfully) by such tricks as "coarse-graining" [2,8]; an *ad hoc* addition of randomness to the system in the form of a postulate of "molecular chaos" or "random phases" [2,5]; or by taking the thermodynamic limit [9,10]. It will be argued in Section V that the results of this paper may make such subterfuges unnecessary, and that there is a deep connection between the increase of gravitational field entropy—which does not seem to require the above tricks—and the increase of matter entropy. The paper will conclude with a discussion of the significance of the no-return theorem for cosmology.

II. Brief History of the Eternal Return Idea

The concept of the eternal return—the idea that time is fundamentally cyclic—apparently played a key role in the cosmological thought of mankind as far back as 6500 B.C. [11, p. 332]. Scientific thought in this period was based on such common sense phenomena as the cycle of the seasons, the rhythm of human life from birth to adulthood to death, and the numerous periodicities in the heavens such as the phases of the moon, the annual motion of the sun through the constellations, and the periodic near return of the planets (the planets were considered gods; see [11, p. 4]) to previous positions in the sky. Under these circumstances a cyclic notion of time is more natural than rectilinear time, and it was cyclic time that dominated the thought of the so-called primitive peoples [12–14].

The early agricultural civilizations—Sumer, Babylon (see [12, p. 44] and [15, p. 360]), Indian [15, p. 353], Mayan [13, p. 88], and Shang-Chou [15, p. 358]—retained and elaborated the notion of cyclic time. The Babylonians, for example, based their concept of time on the periodicities of the planets. In their view, the lifetime of the Universe, or Great Year, lasts about 432,000 yr. The summer of this Great Year would be marked by the conjunction of all the planets in Cancer, and would be accompanied by a universal conflagration; the winter would occur when all planets have a conjunction in Capricorn, and this would result in a universal flood. The cycle then repeats and in some accounts the next cycle is an exact reproduction of the preceding ones [15, p. 362]. The ancient Indians (Hindus, Buddhists, Jains) extended this basic structure of a single Great Year into an entire hierarchy of Great Years. For instance, a destruction and re-creation of individual forms and creatures (but not of the basic substance of the world) occurred every *Kalpa* or day of Brahma. Each day of Brahma had a duration of about 4 billion yr. The elements themselves together with all forms undergo a dissolution into Pure Spirit, which then incarnates itself back into matter every lifetime of Brahma, or about 311×10^{12} yr. (see [15, p. 363] and [16, p. 354]). The Brahma lifetime is the longest cycle in the Indian system, and the cycle is repeated *ad infinitum*.

Among the Greeks, the Stoics were the most fervent believers in the eternal return. They held [14, p. 47] that all objects in the universe were bound together in an absolutely determinate web of actions and reactions, and that this determinism led to a precise recurrence of *all* events. That is, no event is unique and occurs once and for all (for example, the condemnation and death of Socrates) but rather every event has occurred, occurs, and will occur, perpetually; the same individuals have appeared, appear, and re-appear in every return of the cycle. Cosmic duration is thus repetition and *anakuklosis*, eternal return [13, p. 89].

This Stoic idea of *palingenesia*—that is, the reappearance of the same men in each cycle [12, p. 47], carried the idea of the eternal return to its logical extreme, and went much further than the above-mentioned earlier thinkers were willing to go, or indeed Aristotle and Plato were willing to go.

Aristotle was intrigued by the notion of palingenesia. He noted that if it were true it would obscure the usual idea of before and after, for it would imply (see [14, p. 46] and [17]) that he himself was living as much before the fall of Troy as after, since the Trojan War would be reenacted and Troy would fall again. However, although he accepted the cycles, he was reluctant to accept the exact identity of events in each cycle, arguing that the identity was only one of a kind [12, p. 48]. Plato's cosmology was also cyclic, with a periodic destruction and recreation of the universe in conjunction with various astronomical events [18,19]. Indeed, it is through Plato's writings that the notion of the Great Year entered later Western thought. However, scholars disagree as to whether his concept of the cycles went to the extreme of palingenesia (see [12, p. 48] and [14, p. 45]). The eternal return concept came to dominate thought in the pre-Christian, later Roman empire. It was also prominent on the other side of the oikoumene in this period, namely, in Han China. As Needham has pointed out [20, p. 29], the popular religious Taoism of the Han was millenniarist and apocalyptic; the Great Peace was clearly in the future as well as the past. In the Canon of the Great Peace (written between 400 B.C. and 200 A.D.) is found a theory of cycles that issued fresh from chaos (i.e., a state of complete undifferentiation, which is similar to modern ideas of maximal entropy) and then fell slowly until the day of doom (maximal entropy in another guise).

As the epigram to this paper suggests, the Christian world view was hostile to the eternal return idea; in the section of the "City of God" from which the selection quoted is taken, St. Augustine was criticizing the Stoic recurrence concept, arguing that Christian philosophy (and its Hebrew predecessor) required a noncyclic *linear* concept of Time. God created the world *once*, Christ died *once*, and will rise again *once*. With the triumph of Christianity, the notion of linear time became dominant over cyclic time in the West until the rise of modern science, though a few Medieval scholars, such as Bartholomaeus Anglicus (1230), Siger of Bruhant (1270), and Pietro d'Acono (1300), were willing to at least entertain the notion of an eternal return [21]. In Medieval China, however, the Neo-Confucian school, which flourished in the 11th to 13th centuries of our era and which was influenced both by Buddhist ideas on recurrence and the above-mentioned ideas of ancient Taoism, accepted the idea that the universe passed through alternating cycles of construction and dissolution [20, pp. 6, 22]. For instance, the Sung scholar Shen Kua (f. 1050) discussed recurrent world catastrophes (see [20, pp. 22] and [22, pp. 598 *ff.*, 603 *ff.*]), and later the Ming scholar

Tung Ku held that a world period had a beginning, but not the endless chain of all world periods (see [20, p. 6] and [22, p. 406]). Needham has argued [20, p. 50] that in fact the linear notion of time dominated over the cyclic view in later Chinese thought, but other scholars disagree [23,47]. However, there is no question that linearity dominated in Christian thought, and many scholars (e.g. [16,24]) have contended that this notion of time played a key role in the rise of modern science.

Modern science in turn led to a revival of cyclic time. The Newtonian world picture contained both cyclic and linear aspects from the very beginning. Newton himself was worried that his solar system model—based on a linear (mathematical) time—was gravitationally unstable in the long run, and to compensate for this instability he suggested a cyclic process whereby the planets would be replaced as they were periodically perturbed from the orbits by the gravitational action of other bodies [25]. Euler, Laplace, Lagrange, and others had shown by the beginning of the 19th century that the solar system was in fact stable to first order, the gravitational perturbations leading merely to a *cyclic* oscillation of the planetary orbits. However, at about this time the debate on the question of cyclic versus linear shifted from astronomy to geology and thermodynamics [5, p. 553]. The problem in geology was whether the internal heat of the earth could power geological cycles indefinitely, or whether the earth would eventually cool to a "state of Ice and Death," as J. Murray [26] put it in 1814. This question in part stimulated research in thermodynamics [5, Section 14]; by the end of the 19th century, Kelvin and others (see [5] for a detailed history) concluded on the basis of the newly formulated second law of thermodynamics that the heat death was inevitable, and hence a cyclic notion of time was refuted. Kelvin [27] and Tait [28] indeed inferred that the second law implied a creation of the Universe.

Other physicists were unwilling to grant such unlimited validity to the second law, arguing that the creation of the Universe would violate the first law of thermodynamics [29]. Thus somehow the energy dissipated by thermodynamics processes must be periodically reconcentrated into usable form. Rankine, for instance, suggested [30] that heat radiated into space would reach a sort of "ether wall" a finite distance from the earth, at which the radiant heat would be totally reflected and reconcentrated into various "foci." (Rankine's notion of "ether wall" is strikingly similar to the idea of "domain boundary" which arises in spontaneous symmetry breaking in gauge theories [31,32]). Thus the history of the Universe would be cyclic in the long term.

Rankine was basically trying to show that mechanics and the second law were inconsistent. This was first shown in 1890 by Poincaré in his famous recurrence theorem [33] mentioned above. In its most general form

Poincaré's theorem can be proven in any space X on which there is a one parameter map T_t from the sets $\{U\}$ of X into $\{U\}$ and a measure μ on X such that: (1) $\mu(X) = 1$, and (2) $\mu(T_{t_0}(U)) = \mu(T_{t_0 + t}(U))$ for any $U \subset X$ and any t_0, t. In the application to classical mechanics, (1) is assured by requiring the space X to be the phase space of a finite energy mechanical system in a finite box. If μ is the density function ρ in phase space and T_t is the evolution operator for the mechanical system (which is assumed to be Hamiltonian), (2) then follows from Liouville's theorem: $d\rho/dt = 0$. Thus the classical mechanics of a finite system *is* inconsistent with the second law; by Poincaré's theorem, almost all such systems must return arbitrarily closely and infinitely often to almost all previous initial states.

At about the same time Poincaré was developing his theorem, the English philosopher H. Spencer [34] and the German philosopher F. Nietzsche were attempting to make a scientific-sounding argument for the eternal return. Nietzsche's argument is worth repeating in detail because although it is nonrigorous (to say the least!), it contains all the essential ideas that are needed for a rigorous proof (given certain assumptions about the evolution of the world system), a proof which will be useful in discussing the significance of the theorem stated in the next section. I shall follow the example of Nietzsche's sister and omit those parts of his argument which I think are nonsense.

Nietzsche began his "proof" of eternal recurrence as follows:

> ... we insist upon the fact that the world as a sum of energy must not be regarded as unlimited—we forbid ourselves the concept infinite energy, because it seems incompatible with the concept energy [35, #5].

This is similar to the idea of energy in general relativity. Only in asymptotically flat space where the total energy is necessarily finite does the concept energy have a well-defined meaning [37, p. 457]. Nietzsche also argued that the universe must be infinite in time:

> We need not concern ourselves for one instant with the hypothesis of a created world. The concept "create" is today utterly indefinable and unrealisable; it is but a word which hails from superstitious ages ... [36, #1066].

He then contended that recurrence of all states followed from the finiteness of energy (and space) by which he meant a finite number of possible states of the universe, the infinity of elapsed time, and a chancelike evolution:

> If the Universe may be conceived as a definite quantity of energy, as a definite number of centers of energy,— and every other concept remains indefinite and therefore useless, it follows therefrom that the Universe must go through a calculable number of combinations in the great game of chance which constitutes its existence. In infinity, at some moment or

other, every possible combination must once have been realized; not only this, but it must once have been realized an infinite number of times ... [36, #1066]. If all possible combinations and relations of forces had not already been exhausted, than an infinity would not yet lie behind us. Now since infinite time must be assumed, no fresh possibility can exist and everything must have appeared already, and moreover an infinite number of times [35, #7]. That a state of equilibrium has never been reached, proves that it is impossible. [But it must have been reached] in spherical space [36, #1064]. Only when we falsely assume that space is unlimited, and that therefore energy gradually becomes dissipated, can the final state be an unproductive and lifeless one [35, #8].

In the last section of this paper, it will be pointed out that Nietzsche's world model can be compared fairly closely to a Markov process whose states must recur. Thus, granting Nietzsche's assumptions (finite number of states, no creation, and chancelike evolution), his proof of recurrence is valid when judged by the standards of philosophical (not mathematical) rigor.

Another 19th century thinker who considered the problem of recurrence was Boltzmann. Boltzmann originally hoped to deduce irreversibility from the mechanics of atoms but he soon realized such a deduction was impossible without using averaging techniques. Under pressure from Planck's student Zermelo, who based his arguments on Poincaré's theorem, Boltzmann suggested that the Universe as a whole had no time direction, but individual regions of it did, when by chance a large fluctuation from equilibrium should produce a region with reduced entropy. These reduced entropy regions would then evolve back to the more probable state of maximum entropy, and the process would repeat in accordance with Poincaré's theorem (see [5, Section 14.7] and [38]).

Once it became clear that a finite system of *particles* would be recurrent and not irreversible in the long run, Planck considered whether irreversibility could arise from a *field* theory such as electromagnetism. The idea would be to derive irreversibility from the interaction of a continuous field with the discrete particles. Planck began a series of papers on this question in 1897, a series that culminated in his discovery of the quantum theory of radiation in 1900. Boltzmann, however, pointed out that if we regard the field as a system with an infinite number of degrees of freedom, this would be analogous to a mechanical system with an infinite number of molecules, and in either case we would have an infinite Poincaré recurrence time; in either case we would have irreversibility and long term agreement with the second law. However, for the thermodynamics of fields in a confined space, it is physically more appropriate to regard the field not as a continuous quality governed by differential equations, but rather as a large but finite number of "vector ether atoms" whose equations of motion are obtained by replacing the usual differential equations with finite difference equations. To this system the recurrence theorem would apply (see [5, Section 14.8] and [39]).

Most 20th century discussions of the eternal return are based on the so-called oscillating closed universe model developed in 1922 by A. Friedmann [40]. Friedmann himself was aware of the cyclic nature of time in his solution, and suggested that one could identify corresponding times in each cycle. However, in the Friedmann model the radius of the Universe went to zero at the beginning and the end of each cycle, and thus from a strict mathematical standpoint the cycles were disjoined by a singularity; they were not true "cycles." Tolman proved [41] in 1931 that such a discontinuity was inevitable at the beginning and end of any isotropic and homogeneous closed universe with a physically reasonable matter tensor. He argued [42] that this discontinuity was merely an artifact of the high symmetry assumed, and in a physically realistic universe, the actual discontinuity would disappear. He therefore assumed that the entropy would be conserved in a passage through the singularity, and thus the thermodynamics of a cycle would in part be determined by the history of a previous cycle. Other relativists of the time by and large agreed with Tolman as to the unreality of the singularity (see [43] for more details).

With the advent of the Hawking–Penrose singularity theorems in the 1960s, most relativists have accepted the reality of some sort of initial singularity—at least in classical relativity. Some relativists have argued that quantum effects could cause the universe to "bounce" at very high densities, leading to cycles in the closed universe models. Wheeler, for example, has until recently suggested [44,48] that the physical constants themselves are recycled at a "bounce." (Wheeler is presently advocating a one-cycle closed universe. See, e.g. [48]. In acknowledgement of Wheeler's change of view, I have elsewhere [70] termed a one-cycle closed universe a "Wheeler Universe.") Thus Tolman's concept of the cycles was analogous to the "day of Brahma" cycle in Indian mythology, while Wheeler's earlier concept resembled the "life of Brahma" cycle. I shall now show that in classical relativity, the singularity prevents recurrence, and in Section V I shall argue that if quantum effects do result in a "bounce," then some sort of recurrence is probably inevitable.

III. The No-Return Theorem

In order to prove that no two states of the Universe can be identical or even arbitrarily close, we must first make precise the notion of "close." This can be done by regarding the set of all initial data as a *Sobolev space* W^s. The global topology of the space–time (M, g) is $S \times \mathbb{R}$, where S is compact, if the space–time is globally hyperbolic. We choose some positive definite metric e_{ab} on M, and define a norm on W^s by

$$\|K_J^I\|_m \equiv \left[\sum_{p=0}^m \int_S (|D^p K_J^I|)^2 \, d\sigma \right]^{1/2}, \tag{1}$$

$$\|h, \chi\| \equiv \|h\| + \|\chi\|,$$

where $d\sigma$ is the volume element induced on S by e_{ab}, D^p is the generalized pth covariant derivative with respect to some chosen background metric $\bar{g}_{ab}$, $|\ \ |$ is the norm induced by e_{ab} (the metric e_{ab} is usually assumed to be uniform (unvarying) in the $\mathbb{R}$ direction), and K_J^I, h, χ are any tensors (see [45, pp. 233–235] for more details). Two tensors will be "close" if they are "close" in the norm (1).

The no-return theorem will require four conditions on the matter tensor. The first condition, the timelike convergence condition [45, p. 95], says roughly that gravitation is always an attractive force. The other three conditions, which will here be called (a)–(c), are precisely stated by Hawking and Ellis [45, pp. 254–255]. Roughly stated, (a) says that the development from a given set of initial data is locally unique; (b) says that this unique development is locally stable; and (c) says that the stress-energy tensor is a polynomial in the matter fields, their first derivatives, and the space–time metric. As discussed by Hawking and Ellis [45, p. 255], these are physically reasonable conditions to impose on the matter fields, though strictly speaking, they are not necessary conditions. Any other conditions which imply global Cauchy stability and uniqueness (in the sense defined in [45, Chapter 7]) would be sufficient to prove the Theorem below.

We shall also need two global conditions. The first, the requirement of unique development from Cauchy surfaces [45, p. 205], says that Laplacean determinism holds on the space–time. (It is interesting that in general relativity, determinism implies no recurrence, whereas in the Stoic philosophy and classical mechanics, determinism was held to imply recurrence.) The second, the generic condition [45, p. 101] says that every causal geodesic feels a tidal force at least once in its history. As this would be expected to be false only for a "measure-zero" set of solutions to the Einstein equations, the generic condition in the theorem below is somewhat, analogous to the "measure-zero" qualifications in the Poincaré recurrence theorem. A *closed universe* is defined to be a space–time in which the Cauchy surfaces are compact. A space–time that contains two disjoint spacelike Cauchy surfaces that are isometric in their initial data is said to be *time periodic*.

Theorem   If a space–time (M, g) containing compact Cauchy surfaces is uniquely developed from initial data on any of its Cauchy surfaces, and if (M, g) also satisfies both the generic and the timelike convergence conditions, *then the spacetime cannot be time periodic. Furthermore, if in addition the matter fields ψ and their first derivatives ψ' satisfy conditions (a)–(c), then for any neighborhood U of any Cauchy surface S_1, there exists a number $\varepsilon > 0$ such that $\|(h, \chi, \psi, \psi') - (h_1, \chi_1, \psi_1, \psi_1')\|_{5+a} > \varepsilon$ for the initial data on any Cauchy surface S with $U \cap S$ empty. (ε depends on e_{ab}, $\bar{g}_{ab}$, and U. S and S_1 are assumed spacelike.) The tensors h, χ are the three-metric of S and the extrinsic curvature of S, respectively. As discussed in Chapter 7 of [45], (h, χ, ψ, ψ') forms a complete set of initial data for the Cauchy problem in general relativity.*

It should be emphasized that the compactness of the Cauchy surfaces—the requirement of a closed universe—is an essential condition. If the word "compact" is removed from the above theorem, it is false; the theorem does not apply to open universes. For example, a static star in asymptotically flat space satisfies all the conditions of the theorem except compactness, and this spacetime can be foliated by Cauchy surfaces normal to the timelike Killing vector field. The initial data are the same on each of these hypersurfaces.

IV. Proof of the No-Return Theorem

Assume on the contrary that (M, g) is time periodic. By the assumption of unique development, this implies there exists a sequence of Cauchy surfaces $S(t_{-n}), \ldots, S(t_0), \ldots, S(t_n)$ with the initial data on a given surface isometric to the initial data on any other surface, and with $J^+(S(t_i)) \cap J^-(S(t_{i+1}))$ isometric to $J^+(S(t_j)) \cap J^-(S(t_{j+1}))$ for any i, j. Thus (M, g) can be regarded as a covering space for a space–time with topology $S \times S^1$. By the corollary [45, p. 217], there will exist a timelike curve γ_{ij} of maximal length between any two of the isometric Cauchy surfaces $S(t_i)$ and $S(t_j)$, and γ_{ij} will be orthogonal to both. Consider the sequence of curves $\gamma_{-1,1}$, $\gamma_{-2,2}, \ldots, \gamma_{-n,n}, \ldots$. Because $S(t_0)$ is a Cauchy surface, each of these curves intersects $S(t_0)$ in exactly one point. Since $S(t_0)$ is compact, the sequence has a subsequence γ_n that converges to a timelike geodesic γ. [Timelike since in the space $S \times S^1$, we can define γ by a sequence of vectors normal to $S(t_0)$— the vector at the future endpoint of γ_n and tangent to γ_n there. This sequence of vectors has a subsequence that converges to a normal of $S(t_0)$, and all convergent subsequences converge to a normal to $S(t_0)$.]

The geodesic γ is past and future complete. To see this, we first show that the lengths of the geodesic segments γ_n must diverge in both the past and future directions as $n \to \infty$. For since γ_n is the maximal length geodesic segment between $S(t_{-n})$ and $S(t_n)$, and if γ_n converged to a finite length in either direction, the future direction say, then we could construct a causal curve between $S(t_{-n})$ and $S(t_n)$ of length greater than γ_n for n sufficiently large as follows. Every timelike geodesic normal to $S(t_0)$ reaches all $S(t_n)$ since these are Cauchy surfaces. Define a timelike curve $\alpha_n(p)$ from any point p in $S(t_0)$ by extending the geodesic normal to $S(t_0)$ at p until it reaches $S(t_1)$ at p_1, then move along the geodesic normal to $S(t_1)$ at p_1 until this geodesic reaches $S(t_2)$, and so on until $S(t_n)$ is reached. Since $S(t_0)$ is compact, the length of a geodesic from $S(t_0)$ to $S(t_1)$ along a geodesic normal to $S(t_0)$ is bounded below by some number L. Thus the length of $\alpha_n(p) \geq nL$, and so if the length of γ_n did not diverge in the future direction as $n \to \infty$, we could replace γ_n by $[\gamma_n \cap J^-(S(t_0))] \cup [\alpha_n(p = \{\gamma_n \cap S(t_0)\})]$ to obtain a causal curve of length greater than γ_n between $S(t_{-n})$ and $S(t_n)$ for n sufficiently large. But this is impossible, since by definition γ_n is the maximal length curve between these two Cauchy surfaces. By the continuity of length along arbitrarily close continuous geodesic segments, this divergence in both time directions of the lengths of the γ_n segments implies the geodesic completeness of γ.

Since γ is geodesically complete, and since the generic and the timelike convergence conditions hold, γ must have a pair of conjugate points—at points p and q, say—by Proposition 4.4.2 (in [45]). By Proposition 7.24 (in [46]), the location of first conjugate points varies continuously with the geodesic, and so there will be points p_n, q_n on γ_n, which are conjugate points on γ_n and converge to p, q. For n sufficiently large, p_n, q_n will be in $J^+(S(t_{-n})) \cap J^-(S(t_n))$. But γ_n is the maximal length causal curve between $S(t_{-n})$ and $S(t_n)$, and so by Proposition 4.5.8 (in [45]), it cannot have conjugate points in $J^+(S(t_{-n})) \cap J^-(S(t_n))$. (We could also obtain a contradiction by arguing as in [45, p. 270].) This contradiction shows that time periodic spacetimes do not exist.

We shall now show that it is also not possible to approach arbitrarily closely to a previous initial state. If there is no such ε for some S with initial data (h, χ, ψ, ψ'), then there will exist a sequence of Cauchy surfaces S_n with initial data $(h_n, \chi_n, \psi_n, \psi'_n)$ such that $(h_n, \chi_n, \psi_n, \psi'_n) \to (h, \chi, \psi, \psi')$ as $n \to \infty$, with $S_n \cap U$ empty for all n. We may assume without loss of generality that either $S_n \subset I^+(S)$ or $S_n \subset I^-(S)$ for all n. Suppose $S_n \subset I^+(S)$. By (a)-(c), the Cauchy stability theorem [45, pp. 253, 255] holds, and this means that for n sufficiently large ($n > n_1$, say), there is a compact four-dimensional region V such that the minimum length in V of all timelike geodesics normal to S_n is greater than some positive number c, independent of n, provided $n > n_1$. This implies that we can find an infinite sequence of Cauchy surfaces $\tilde{S}_n$ in $I^+(S)$ such that the minimum length from $\tilde{S}_{n-1}$ to $\tilde{S}_n$ along the timelike geodesic normal to $\tilde{S}_n$ is greater than c, and such that the initial data on each $\tilde{S}_n$ is arbitrarily close to the data on S. (If the S_n approach a limit in M, which we can call $\tilde{S}$, then Cauchy stability implies there must be an infinite sequence of such $\tilde{S}$, and this sequence will give the $\tilde{S}_n$. If the S_n do not approach an $\tilde{S}$ in M, then there will be a subsequence of the S_n, which will be the $\tilde{S}_n$.) Again by Cauchy stability, there will be a similar sequence $\tilde{S}_{-n}$ in $I^-(S)$. With these sequences of Cauchy surfaces $\tilde{S}_n$ and $\tilde{S}_{-n}$, we can proceed as in the time periodic case to obtain a contradiction.

V. Significance of the No-Return Theorem

One might think [68,69] that any field theory in Euclidean space would have a nonrecurrence property similar to the one demonstrated in Section IV, since a continuous field has an infinite number of degrees of freedom. From a physical point of view, this is not the case. It is not possible to make a precise measurement of the field variables at every point, and in practice a field restricted to a finite region S would be approximated by dividing up S into a finite number of subregions, and the field in each subregion replaced by its average value in that subregion. Evolution would be via the differential field equations, but one would compare the "average" values.

Comparing initial data sets via the Sobolev norm is closely analogous to comparing the average values of the field variables (and their derivatives) at one time with the average values at another time. With the Sobolev norm, one essentially takes the absolute square average value of the initial data over the entire Cauchy surface rather than dividing up the Cauchy surface into subregions; the Sobolev average is a coarser average. The arbitrary metrics e_{ab} and $\bar{g}_{ab}$ in the Sobolev norm are analogous to an arbitrary coordinate system with respect to which averages are calculated. In the classical mechanics of fields in a finite box, one can define a similar norm $\| \quad \| = [\int |\psi|^2 \, d\sigma]^{1/2}$ on the classical field ψ. The sum $\sum_{i=1}^{n} \|\psi^{(i)}\|$ with $\psi^{(i)}$ the ith derivative of ψ, is essentially the same as the Sobolev norm. This sum is a map f from the space of all values of the field and its first n derivatives into $\mathbb{R}$. If the range of f is bounded—as it must be if the solutions ψ are stable—then the evolution of ψ must result in accumulation points in the range of f. An analogous accumulation is impossible in general relativity, as the above theorem shows. What happens in general relativity is that the range of f, so to speak, is not bounded—singularities develop in spacetime (this follows from the hypotheses of the above theorem and Theorem 2 in [45, p. 266].) This sort of singular behavior is typical of many nonlinear field theories [59,60], but in linear field theories such as electromagnetism, the solutions are stable, hence bounded. For example, the solutions to the Schrödinger equation in a finite box must recur [53,54] in the above average sense: for any initial value $\psi(t_0)$ for the probability distribution and any ε, there exists a time t for which $\|\psi(t) - \psi(t_0)\| < \varepsilon$.

Another argument for field theory recurrence in an average sense is Boltzmann's method of replacing the differential field equations by finite difference equations. As Boltzmann pointed out, the solutions of these difference equations would have a recurrence property.

This leads one to suspect that it might be possible to prove a Poincaré theorem for the average values of linear Hamiltonian classical fields and their first n derivatives. Unfortunately, this will not be easy to do because of the difficulty of defining the required invariant measure on the initial data space. It is known [51], for instance, that a translation invariant measure on a Hilbert space cannot give a finite value on a box in the Hilbert space. The best result to date on this problem is the definition of an infinitesimally invariant measure on the solution space to the two-dimensional Euler equation [50].

In a finite universe without singularities, there should be a finite number of physically distinguishable states (this should also be true if the matter in the universe is in the form of fields, for reasons stated at the beginning of this section). I would expect that the states of such a universe would evolve in general in a quasi-ergodic way [56,67], though it is impossible to prove this without a suitable measure. If the evolution of the physically distinguishable states is indeed quasi-ergodic, then in infinite time, every state would with high probability occur an infinite number of times.

Another way to model the evolution of such a finite state, always existing universe is by regarding the evolution as a finite discrete Markov chain [49] with stationary transition probabilities. That is, we shall sample the state of the universe at definite time intervals Δt, and we shall assume that the state of the universe at time t_j is determined by the previous state at time t_{j-1} and the probability matrix of going from this previous state to any other state. If the universe has existed for an infinite time, then with probability one the states of the universe form a closed set [49, p. 384]; that is, we may consider without loss of generality the Markov chain representing the universe now to be irreducible (This is essentially Nietzsche's argument that if the universe had a final state, it would have reached it by now. It is also Birkhoff's postulate of metric transitivity [55,67].) By Theorem 4 in Feller [49, p. 392], it follows that with probability one, all states recur in the future. Thus if Nietzsche's "chancelike" evolution is assumed to be Markovian, then his argument for recurrence is valid.

Similar arguments will imply some form of recurrence in closed universes that avoid singularities by "bouncing" at some small radius (such behavior could of course occur only if some of the conditions in the no-return theorem are violated [61]). One could envisage the physical constants of the universe, such as the elementary particle masses, the coupling constants, the specific entropy per baryon, and so forth, changing in some fashion at each bounce.

These physical constants can be regarded as additional variables in the initial data. If the number of such constants is finite and if their range of variation is finite, then in a finite universe one would expect a finite number of physically distinguishable states. This implies accumulation points in the state space with an infinite number of bounces; with quasi-ergodic or Markovian evolution, we would have recurrence of all states with high probability. Quantum mechanical considerations do not substantially alter this conclusion, so long as one requires the number of physically distinguishable states to be finite.

It is of course possible to avoid the recurrence conclusion by assuming that the range of variation of the physical constants is not bounded, or that the universe, although closed, increases its radius at maximum expansion with each bounce. (Tolman [41], for example, argued that the monotone increase of entropy required a monotone increase of maximum radius at each bounce.) However, if the range of physical constant variation is not bounded, then there would exist a sequence of cycles such that at least one physical constant would diverge in the limit. I would regard such a divergence as a singularity "at infinity." If the maximum radius increases monotonically with time, then with an infinite number of bounces in the past, the sequences of values of the maximum radius must have a limit point in the past. If its limiting value is nonzero, then one must have recurrence in the past, though this "recurrence" may take the form of a gradual cessation of change, as in the Eddington–Lemaitre universe [43]. If its limiting value is zero, then the singularity has not really been removed, it has just been placed at temporal infinity.

But the singularities in the single cycle closed universe can also be regarded as at temporal infinity. In fact, both Milne [62] and Misner [63] have contended that because physical changes occur with diverging rapidity (as measured in proper time) near a singularity, the singularity should be regarded as occurring at temporal infinity as measured in physical time. Barrow and myself [57] have pointed out that extrinsic time [64], which is a naturally defined absolute time in closed universes [65], has just this property of placing the singularities at temporal infinity. In the extrinsic time scale, a closed universe has existed and will exist forever; the question of what preceded the Big Bang does not arise.

Penrose has suggested ([56]; see however [58]) that the entropy of the gravitational field is proportional to "some suitable integrated measure of the size of the Weyl curvature," and that this curvature is zero at the initial singularity, and infinite at the final singularity. There are indications (e.g. [57]) that in stable solutions with the Weyl curvature initially zero, the average Weyl curvature would increase monotonically with time to the final singularity. If this is indeed the case, the Penrose gravitational entropy could be shown to increase from the initial to the final singularity without the use of "coarse graining," and thus could act as the ultimate source of all forms of entropy increase.

Because of Poincaré recurrence, it is impossible to define a monotonically increasing entropy (or any monotonically increasing function) for non-gravitational fields or systems of particles in terms of the phase space variables of the fields. However, since Poincaré recurrence does not hold for gravitational fields, it *is* possible to define functions that increase monotonically with time—such functions are called Lyapounov functions [9,10]—on the phase space of the gravitational field. In fact, the extrinsic time, which is proportional to the trace of the extrinsic curvature of a leaf of a constant mean curvature Cauchy hypersurface foliation of spacetime, is just such a function since the extrinsic curvature is essentially the momentum of the gravitational field [37], and the extrinsic time increases monotonically in closed universes, ranging from $-\infty$ at the initial singularity to $+\infty$ at the final singularity [65]. In nonsingular asymptotically flat space, a foliation of constant mean curvature Cauchy surfaces would have to have extrinsic time constant $(=0)$. This supports the conclusion of classical mechanics that isolated systems cannot define a time direction [52].

Niels Bohr [66] together with Prigogine and co-workers [6,10] have suggested on the basis of Poincaré recurrence that thermodynamic concepts such as temperature and entropy are complementary to the phase space variables. In other words, they argue that a detailed microscopic description of the behavior of a physical system would preclude the definition of thermodynamic variables for that system. If the gravitational field (or more precisely, the global structure of space–time) is taken into account, we see that it becomes possible in principle to define the thermodynamic variables and the phase space variables in a closed universe simultaneously; the notion of complementarity becomes unnecessary. This leads one to speculate about the connection between quantum mechanics and the global structure of space–time: perhaps the impossibility of giving a detailed deterministic microscopic description of a quantum mechanical system is due to neglect of the global structure of space–time.

Acknowledgments

I am grateful to J. D. Barrow, S. G. Brush, P. R. Chernoff, R. O. Hansen, O. E. Lanford, L. Lindblom, J. E. Marsden, J. Needham, D. W. Sciama, N. Siven, and J. A. Wheeler for helpful discussions.

References

[1] Kac, M., "Probability and Related Topics in Physical Sciences." Wiley (Interscience), New York, 1959.
[2] Davies, P. C. W., "The Physics of Time Asymmetry." Univ. of California Press, Berkeley, 1975.

[3]   Holton, G., "Thematic Origins of Scientific Thought: Kepler to Einstein." Harvard Univ. Press, Cambridge, Massachusetts, 1973.

[4]   Holton, G., *Science* **188**, 328 (1975); Merton, R. K., *Science* **188**, 335 (1975).

[5]   Brush, S. G., "The Kind of Motion We Call Heat: A History of the Kinetic Theory of Gases in the 19th Century," Vol 2. North-Holland Publ., Amsterdam, 1976.

[6]   Prigogine, I., George, C., Henin, F., and Rosenfeld, L., *Chem. Scr.* **4**, 5 (1973).

[7]   Prigogine, I., Mayné, E., George, C., and de Hann, M., *Proc. Natl. Acad. Sci. U.S.A.* **74**, 4152 (1977).

[8]   Uhlenbeck, G. E., *in* "The Physicist's Conception of Nature" (J. Mehra, ed.), p. 501. Reidel Publ., Dordrecht, Netherlands, 1973.

[9]   Prigogine, I., *Science* **201**, 777 (1978).

[10] Misra, B., *Proc. Natl. Acad. Sci. U.S.A.* **75**, 1627 (1978).

[11] de Santillana, G., and von Dechend, H., "Hamlet's Mill: An Essay on Myth and the Frame of Time." Gambit, Boston, Massachusetts, 1969.

[12] Baillie, J., "The Belief in Progress," p. 42. Oxford Univ. Press, London and New York, 1950.

[13] Eliade, M., "The Myth of the Eternal Return." Pantheon, New York, 1954.

[14] Toulmin, S., and Goodfield, J., "The Discovery of Time." Harper, New York, 1965.

[15] Sorokin, P. A., "Social and Cultural Dynamics," Vol. 2. Amer. Book Co., New York, 1937.

[16] Jaki, S., "Sciences and Creation." Scottish Academic Press, Edinburgh, 1974.

[17] Aristotle, *Problemata*, Book XVII, 3.

[18] Plato, *Timaeus*, 39.

[19] Plato, *Politicus*, 269C.

[20] Needham, J., "Time and Eastern Man," Occasional Paper, No. 21. R. Anthropol. Inst. London, 1965.

[21] Thorndike, L., "A History of Magic and Experimental Science During the First 13 Centuries of Our Era," Vol. 2, pp. 203, 370, 418, 589, 710, 745, 895. Columbia Univ. Press, New York, 1947.

[22] Needham, J., "Science-and Civilization in China," Vol. 3. Cambridge Univ. Press, London and New York, 1960.

[23] Sivin, N., *Earlham Rev.* **1**, 82 (1966).

[24] Tillich, P., "The Protestant Era." Univ. of Chicago Press, Chicago, Illinois, 1948; see also Cullmann, O., "Christ and Time." Westminister, Philadelphia, Pennsylvania, 1950.

[25] Kubrin, D., *J. Hist. Ideas* **28**, 325 (1967).

[26] Murray, J., *Trans. R. Soc. Edinburgh* **7**, 411 (1815).

[27] Thompson, S. P., "The Life of William Thomson, Baron Kelvin of Largs," p. 111. Macmillan, London, 1910.

[28] Brush, S. G., *Grad. J.* **7**, 477 (1967), Footnote 32 (p. 547).

[29] Arrhenius, S., "Worlds in the Making: The Evolution of the Universe," p. 193. Harper, New York, 1908.

[30] Rankine, W. J. M., *Philos. Mag.* **4**, 358 (1858).

[31] Weinberg, S., "The First Three Minutes," p. 144. Basic Books, New York, 1977.

[32] Everett, A. E., *Phys. Rev. D* **10**, 3161 (1974).

[33] Poincaré, H., *Acta Math.* **13**, 1 (1890); Engl. transl. in Brush, S. G., "Kinetic Theory. Vol. 2: Irreversible Processes." Pergamon, Oxford, 1966.

[34] Spencer, H., "First Principles," 6th ed., Chap. 23. Appleton, London, 1910.

[35] Nietzsche, F., Eternal recurrence, *in* "Complete Works of Friedrich Nietzsche" (O. Levy, ed.), Vol. 16. Foulis, Edinburgh, 1910.

[36] Nietzsche, F., The will to power, Vol. 2, Book 4, *in* "Complete Works of Friedrich Nietzsche" (O. Levy, ed.), Vol. 15. Foulis, Edinburgh, 1910.

[37] Misner, C. W., Thorne, K. S., and Wheeler, J. A., "Gravitation." Freeman, San Francisco, California, 1973.
[38] Boltzmann, L., *Ann. Phys. (Leipzig)* **59**, 793 (1896); Engl. transl. in Brush, S. G., "Kinetic Theory. Vol. 2: Irreversible Processes," p. 238. Pergamon, Oxford, 1966.
[39] Boltzmann, L., *Sitzungsber. Preuss. Akad. Wiss. Berlin*, p. 1016 (1897).
[40] Friedmann, A., *Z. Phys.* **10**, 377 (1922).
[41] Tolman, R. C., "Relativity, Thermodynamics and Cosmology." Oxford Univ. Press, London, 1934.
[42] Tolman, R. C., *Phys. Rev.* **38**, 1758 (1931).
[43] Tipler, F. J., Clarke, C. J. S., and Ellis, G. F. R., *in* "General Relativity and Gravitation: One Hundred Years after the Birth of Albert Einstein" (A. Held, ed.), Vol. 2, p. 97. Plenum, New York, 1980.
[44] Wheeler, J. A., Ref. 37, Chap. 44.
[45] Hawking, S. W., and Ellis, G. F. R., "The Large Scale Structure of Space–Time." Cambridge Univ. Press, London and New York, 1973.
[46] Penrose, R., "Techniques of Differential Topology in Relativity." SIAM, Philadelphia, Pennsylvania, 1972.
[47] Sivin, N., "Cosmos and Computation in Early Chinese Mathematical Astronomy." Brill, Leiden, 1969.
[48] Wheeler, J. A., "Frontiers of Time." North-Holland Publ., Amsterdam, 1979.
[49] Feller, W., "An Introduction to Probability Theory and Its Applications," 3rd ed., Vol. 1. Wiley, New York, 1968.
[50] Albeverio, S., Ribeiro de Faria, M., and Høegh-Krohn, R., "Stationary Measures for the Periodic Euler Flow in Two Dimensions," Prepr. Math. Inst., Univ. of Oslo, Oslo, 1978.
[51] Choquet-Bruhat, Y., DeWitt-Morette, C., and Dillard-Beick, M., "Analysis, Manifolds, and Physics, p. 514. North-Holland Publ., Amsterdam, 1977.
[52] Reichenbach, H., "The Direction of Time." Univ. of California Press, Berkeley, 1971.
[53] Percival, I. C., *J. Math. Phys.* **2**, 235 (1961).
[54] Bocchieri, P., and Loinger, A., *Phys. Rev.* **107**, 337 (1957).
[55] Farquhar, I. E., "Ergodic Theory in Statistical Mechanics." Wiley (Interscience), New York, 1964.
[56] Penrose, R., *in* "Theoretical Principles in Astrophysics and Relativity" (N. R. Lebovitz, W. M. Reid, and P. O. Vandervoort, eds.), p. 217. Univ. of Chicago Press, Chicago, Illinois, 1978.
[57] Barrow, J. D., and Tipler, F. J., *Nature (London)* **276**, 453 (1978).
[58] Gibbons, G. W., and Hawking, S. W., *Phys. Rev. D* **15**, 2738 (1977).
[59] Glassey, R. T., *J. Math. Phys.* **18**, 1794 (1977).
[60] Levine, H., *Trans. Am. Math. Soc.* **192**, 1 (1974).
[61] Tipler, F. J., *Phys. Rev. D* **17**, 2521 (1978).
[62] Milne, E. A., "Modern Cosmology and the Christian Idea of God." Oxford Univ. Press, London, 1952.
[63] Misner, C. W., *Phys. Rev.* **186**, 1328 (1969).
[64] York, J. W., *Phys. Rev. Lett.* **28**, 1082 (1972).
[65] Marsden, J. E., and Tipler, F. J., "Maximal Hypersurfaces and Foliations of Constant Mean Curvature in General Relativity," University of California at Berkeley preprint, 1980.
[66] Bohr, N., *J. Chem. Soc.* p. 349 (1932).
[67] Lebowitz, J. L., and Penrose, O., *Phys. Today* **26**, 23 (1973).
[68] Ehrenfest, P., and Ehrenfest, T., "The Conceptual Foundations of the Statistical Approach in Mechanics," p. 63. Cornell Univ. Press, Ithaca, New York, 1959.
[69] Ref. 2, p. 62.
[70] Tipler, F. J., *Nature (London)* **270**, 500 (1977).

4

Energy and Momentum of the Gravitational Field

James W. York, Jr.

Department of Physics and Astronomy
University of North Carolina
Chapel Hill, North Carolina

Abstract

In this article I review the standard two-surface integral formulation of gravitational energy and momentum and give a detailed discussion of their asymptotic gauge invariances. I state, summarize, and criticize the important attempts to prove the truth of the positive energy conjecture, ending with an assessment of the current status of this work. I point out some difficulties with the concept of angular momentum in general relativity and propose some partial solutions.

I. Introduction

An asymptotically flat space–time should correspond in some reasonable sense to the gravitational field of an "isolated" system. In such cases, one expects the notions of *total* energy, momentum, and angular momentum, measured at large spacelike distances from an origin, i.e., at "spatial infinity," to carry over from field theories in special relativity to general relativity. It is of course well known that the local counterparts of these quantities are not in general well defined in curved space–time. It is also well known that even the concepts of total energy and momentum can only be defined to be zero in closed universes.

In this article, I shall review the standard, easy-to-use, two-surface integral formulation of these concepts and their behavior under asymptotic coordinate transformations. Special attention is devoted to the positive energy conjecture and the difficulties encountered in obtaining a proof

of its validity. I shall also describe the fact that the total angular momentum
J is asymptotically gauge invariant in a weaker sense than the energy E and
linear momentum **P**.

The concepts of "asymptotically flat space–times" and "isolated systems"
have been defined in different ways too numerous to review here. In this paper,
I shall describe the concepts in terms of the Cauchy, or initial, data on a
spacelike hypersurface. The motivation for this approach is that if E, **P**, and **J**
are to be well defined at spatial infinity, they must be conserved in the
evolution of the initial data. Hence they must be expressible entirely in terms
of the initial data and must *not*, in particular, depend explicitly on the lapse
function α or the shift vector β, the choice of which influences the behavior
of the Cauchy data only on nearby slices. (For a review of these quantities
and the modern "3 + 1" description of gravity dynamics and the initial-
value problem, see [1].) In short, E, **P**, and **J** should be expressible in terms
of only the instantaneous state of the gravitational and matter fields.

The appropriate initial data on a spacelike slice Σ are the metric g_{ij} of Σ,
its extrinsic curvature K_{ij}, and the local energy density μ and momentum
density j^i of the matter fields. Let R denote the scalar curvature of g_{ij} and ∇_i
its covariant derivative. Then the initial-value constraints are, with $G =
c = 1$ and standard spacelike conventions [2],

$$R - K_{ij}K^{ij} + (\text{tr}_g K)^2 = 16\pi\mu, \tag{1}$$

$$\nabla_j(K^{ij} - g^{ij}\,\text{tr}_g K) = 8\pi j^i. \tag{2}$$

These are called, respectively, the energy constraint and the momentum
constraint. The quantity $\text{tr}_g K = g_{ij}K^{ij}$ is the mean (extrinsic) curvature of
Σ and will play an important role in what follows.

For the most part, I shall assume that Σ is an $\mathbb{R}^3$ manifold, that is, dif-
feomorphic to $\mathbb{R}^3$. However, the topics to be treated will be described in
such a way as to carry over, for the most part, to regular topologically
nontrivial initial manifolds with asymptotically flat "ends" (e.g., the time-
symmetric slice of the vacuum Schwarzschild–Kruskal black hole) and to
manifolds that may contain singularities, if they are enclosed by apparent
horizons (assuming "cosmic censorship" [3,22] in such cases).

II. Asymptotically Flat Initial Data

A precise context in which to describe asymptotic flatness of Cauchy
data is provided by the "weighted Sobolev spaces" $M^p_{s,\delta}$ introduced and used
in proving a number of important results in general relativity by Cantor [4].

However, here I shall use the usual "order of growth" description. Each such assertion has a corresponding precise statement in $M_{s,\delta}^p$ language.

The purpose of the asymptotic conditions that I shall state is to build in several related properties:

(a) E and $\mathbf{P}$ are finite. ($\mathbf{J}$ will be discussed separately in the final section of this article.)

(b) Solutions of the constraints actually have these properties. (See, e.g. [5].) This point should be emphasized, because most studies of the topics addressed here have not been based on established generic properties of solutions of the constraints.

(c) There are no real or "coordinate" waves too near spatial infinity. The former would presumably make E diverge (which would lead to closure of the spacelike slice [6]); the latter introduces unnecessary complications that we know how to eliminate, in any event [7].

I should mention that if we only require condition (a), the requirements below can be weakened (i.e., slower falloff can be allowed) [8]. However, all of the essential features of our discussion arise without this slight increase in mathematical generality.

I shall assume that r denotes the standard Euclidean distance measured from an arbitrary "origin," which might, in fact, be the topologically spherical boundary of a compact set. For r sufficiently large, we require for our asymptotically flat Cauchy data that

$$g_{ij} - \delta_{ij} = O(r^{-1}); \qquad g_{ij,k} = O(r^{-2}); \qquad g_{ij,kl} = O(r^{-3}); \qquad \text{etc.,} \quad (3)$$

$$K_{ij} = O(r^{-2}); \qquad K_{ij,k} = O(r^{-3}); \qquad \text{etc.,} \quad (4)$$

$$\mu = O(r^{-4}); \qquad j^i = O(r^{-4}). \quad (5)$$

Here asymptotically Cartesian coordinates are used for simplicity; expressing these conditions covariantly is very simple [1]. Covariant descriptions will be used only when they are needed to illuminate certain points. The "etc.s" refer to the fact that, in general, when we state $h_{ij} = O(r^{-1})$, we assume successively faster falloff of each derivative that appears in the calculations, with consequent possible loss of degree of differentiability. The latter is of no importance for present purposes.

What kinds of "coordinate transformations" (usually thought of as "active," i.e., as diffeomorphisms acting by "pull-back") preserve the above conditions? First we consider purely spatial transformations acting within the given slice. We assume from here on that all transformations considered are smooth and have inverses with the same properties as stated for the transformation itself. We shall permit any transformation such that, for large r,

$$x^i \to \bar{x}^i = F^i(x), \tag{6}$$

$$\left(\frac{\partial F^i}{\partial x^j} - \delta^i_j\right) = O(r^{-1}). \tag{7}$$

These include transformations that reduce to the identity at infinity, those which reduce to Euclidean translations at infinity, and those which are of "$O(1)$" at infinity. The latter will be referred to as "asymptotic gauge transformations," e.g., $\bar{x} = x + x^2 r^{-2}$, $\bar{y} = y$, $\bar{z} = z$. Gauge transformations must be considered because of the ambiguity inherent in choosing, for a given asymptotically flat g_{ij}, the flat "background" metric δ_{ij}. To the above, we should also append the Euclidean rotations $\bar{x}^i = A^i_j x^j$, A^i_j is the constant orthogonal rotation matrix. These are of $O(r)$ but are obviously permissible. So, for completeness, (7) should be replaced by

$$\left(\frac{\partial F^i}{\partial x^j} - \delta^i_j\right) = A^i_j + O(r^{-1}). \tag{8}$$

To speak of transformations of the "time coordinate," we suppose that the asymptotically flat data are given on a regular $\mathbb{R}^3$ Cauchy hypersurface and that there is, therefore, a finite maximum domain of dependence [9] and a space–time $\Sigma \times \mathbb{R}$. Singularities are allowed to develop in the future of Σ, but if so, we assume that they are enclosed by an apparent horizon and that cosmic censorship holds.

We extend the above remarks to include time transformations (e.g., moving to a different slice in the given space–time) by requiring (μ, $\nu = 0, 1, 2, 3$)

$$x^\mu \to \bar{x}^\mu = F^\mu(x, t), \tag{9}$$

$$\left(\frac{\partial F^\mu}{\partial x^\nu} - \delta^\mu_\nu\right)\bigg|_\Sigma = O(r^{-1}). \tag{10}$$

The extension to include Lorentz transformations, including Euclidean rotations, replaces the right-hand side of (10) by $\Lambda^\mu_\nu + O(r^{-1})$, with Λ^μ_ν a constant Lorentz matrix:

$$\left(\frac{\partial F^\mu}{\partial x^\nu} - \delta^\mu_\nu\right)\bigg|_\Sigma = \Lambda^\mu_\nu + O(r^{-1}). \tag{11}$$

To use the above transformations in the stated form, we should mention that the spacetime metric is given by

$$ds^2 = g_{ij}\,dx^i\,dx^j + 2\beta_i\,dx^i\,dt - (\alpha^2 - g^{ij}\beta_i\beta_j)dt^2, \tag{12}$$

where $\beta_i \, dx^i$ is a "shift one-form" and α is a lapse function. (Note $g_{ij}g^{jk} = \delta_i^k$, $\beta^i = g^{ij}\beta_j$.) We assume, in parallel to (3)–(5) that

$$\alpha - 1 = O(r^{-1}); \qquad \beta_i = O(r^{-1}). \tag{13}$$

The evolution of the Cauchy data is given by the Einstein equations for $\dot{g}_{ij}$ and $\dot{K}_{ij}$ (see, e.g. [1]), with the dot denoting an evolution vector field on Σ:

$$\cdot = \frac{\partial}{\partial t} = \alpha \frac{\partial}{\partial n} + \beta^i \frac{\partial}{\partial x^i}, \tag{14}$$

where $\partial/\partial n$ is the unit timelike future-directed normal of Σ, to which β^i is orthogonal.

Clearly, the action of an infinitesimal gauge transformation generated by $\partial/\partial\xi$ on the Cauchy data can also be computed from the Einstein equations by replacing $\partial/\partial t$ by $\partial/\partial\xi$, where

$$\frac{\partial}{\partial\xi} = \xi^* \frac{\partial}{\partial n} + \xi^i \frac{\partial}{\partial x^i}, \tag{15}$$

$$\xi^* = O(1); \qquad \xi_i = O(1), \tag{16}$$

$$\xi^*_{,i} = O(r^{-1}); \qquad \xi_{i,j} = O(r^{-1}). \tag{17}$$

When a quantity is unchanged by the application of $\partial/\partial t$, we shall say that it is a constant of the motion (conserved), and, when unchanged by $\partial/\partial\xi$, that it is gauge-invariant. These two notions could be unified, but we prefer to keep them distinct. It is easy to verify that conditions (3)–(5) are preserved under both time evolution and gauge transformations.

III. Energy and Linear Momentum

There are several useful two-surface integrals at spatial infinity that define the energy [10] and do not depend on the lapse and shift functions [11]. A basic one, in Euclidean coordinates, is

$$E = \frac{1}{16\pi} \oint_\infty (g_{ij,j} - g_{jj,i})d^2 S_i, \tag{18}$$

where ∞ denotes the standard Euclidean sphere $r =$ constant, in the limit $r \to \infty$. For both the energy and the momentum, as well as the angular momentum, it is useful to demonstrate gauge invariance without appealing to Gauss's theorem and volume integration, because the interior could in general contain singularities. Also, the initial manifold might be multiply connected with several asymptotically flat "ends." Therefore, we shall stay on the two-sphere and use the theorem of Stokes plus the fact that

quantities falling off faster than r^{-2} do not contribute to the surface integral.
    Recall from Einstein's equations in first-order form that

$$\frac{\partial g_{ij}}{\partial \xi} = -2\xi^* K_{ij} + \nabla_i \xi_j + \nabla_j \xi_i, \tag{19}$$

with $\xi_j \equiv g_{ij}\xi^i$. If we write $g_{ij} = f_{ij} + h_{ij}$, where f_{ij} is the flat "background metric" (in arbitrary coordinates), and express (18) in terms of covariant derivatives of h_{ij}, we have

$$E = \frac{1}{16\pi} \oint_\infty \nabla^j(h^i_j - \delta^i_j \operatorname{tr} h)d^2 S_i. \tag{20}$$

Applying (19) to (20), we find

$$\frac{dE}{d\xi} = \frac{1}{16\pi} \oint_\infty [\nabla_j(\nabla^j\xi^i - \nabla^i\xi^j) + O(r^{-3})]d^2 S_i. \tag{21}$$

Therefore, the "dangerous" ($O(r^{-2})$) term vanishes by Stokes's theorem and $\partial\partial \equiv 0$ ($\partial = $ boundary operator).
    In parallel to (20), the total linear momentum **P** in the direction of a translational Killing vector ϕ^i of the flat metric is given by

$$P^i\phi_i = P_\phi = \frac{1}{8\pi} \oint_\infty (K^i_j - \delta^i_j \operatorname{tr} K)\phi^j \, d^2 S_i. \tag{22}$$

We need the Einstein equation

$$\frac{\partial K_{ij}}{\partial \xi} = -\nabla_i\nabla_j\xi^* + \xi^*[R_{ij} - 2K_{ik}K^k_j + K_{ij} \operatorname{tr} K$$

$$- 8\pi(S_{ij} - \tfrac{1}{2}g_{ij} \operatorname{tr} S) - 4\pi\mu g_{ij}]$$

$$+ \xi^l\nabla_l K_{ij} + K_{il}\nabla_j\xi^l + K_{lj}\nabla_i\xi^l. \tag{23}$$

We assume the (spatial) stress tensor S_{ij} (projection of $T_{\mu\nu}$ onto Σ) has growth $O(r^{-4})$, just as do μ and j^i.
    In computing $dP_\phi/d\xi$, it is convenient to introduce the abbreviation $p_{ij} = \operatorname{tr} Kg_{ij} - K_{ij}$ (so $\pi^{ij} = (\det g)^{1/2}p^{ij}$ is the usual canonical momentum). Also, we consider tangential (ξ^i) and normal (ξ^*) deformations of Σ separately. For the tangential deformations

$$\left(\frac{\partial P_\phi}{d\xi}\right)_{\text{tangt}} = -\frac{1}{8\pi} \oint_\infty \lambda^i \, d^2 S_i, \tag{24}$$

$$\lambda^i = p^i_j \mathscr{L}_\phi \xi^j + \nabla_j[2p^{[i}_k\xi^{j]}\phi^k] + \xi^i[\phi^j\nabla_k p^k_j + 2p^{jk}\nabla_{(j}\phi_{k)}]. \tag{25}$$

Each term falls off at least as fast as $O(r^{-3})$ and we find zero in (24). This expression will be considered again in the case of angular momentum.

For normal deformations, one finds

$$
\left(\frac{dP_\phi}{d\xi}\right)_{norm} = -\frac{1}{8\pi} \oint_\infty [\nabla_j F^{ij} + \nabla_j H^{ij} - \nabla^i \xi^*(\nabla_j \phi^j)
$$

$$
+ \nabla_j(\xi^* \nabla^{(i} \phi^{j)}) - 2\xi^* \nabla_j \nabla^{(i} \phi^{j)} - \xi^* \nabla^i \nabla_j \phi^j
$$

$$
+ O(r^{-4})] \, d^2 S_i, \tag{26}
$$

where

$$
\begin{aligned}
F^{ij} &= F^{[ij]} = \phi^j \nabla^i \xi^* - \phi^i \nabla^j \xi^*, \\
H^{ij} &= H^{[ij]} = \xi^* \nabla^{[i} \phi^{i]}.
\end{aligned} \tag{27}
$$

Because ϕ^i is a translation, the only potentially troublesome term is the one involving F^{ij}, which, however, is disposed of using Stokes's theorem. Therefor **P** is gauge-invariant and conserved.

It is also most important in what follows to know that E and **P** together form a Lorentz four-vector P^μ with $P^0 = E$, $P^i = P^i$ as defined above. This means that P^μ transforms properly under Lorentz boosts ($P^{\mu'} = \Lambda^\mu_\nu P^\nu$, including $P^{i'} = A^i_j P^j$) and is invariant with respect to all other transformations allowed by (11). The Lorentz-covariance was demonstrated by Arnowitt et al. [10], and others [12], and I will not repeat the demonstration here. Of course, ADM also discussed asymptotic gauge invariance, but not with the detailed covariant methods used here. The latter are also helpful in discussing angular momentum.

IV. Boosted Schwarzschild Initial Data

There is a simple example that illuminates the preceding discussion and is quite relevant to recent attempts to settle the positive energy conjecture [13]. Suppose we have a spherical "star" with a Schwarzschild vacuum exterior. Let us look at the asymptotic data for this situation in a boosted reference system that is far away from the star. (Equivalently, we shall obtain the asymptotic data for a star that has been moving at constant speed for a "very long time.") These data will be shown to satisfy the criteria for asymptotic flatness, as expected. The corresponding space–time is certainly asymptotically flat at spatial infinity. (Compare the Maxwell field of a uniformly moving electric charge.)

In isotropic coordinates, the data of the exterior field in the usual frame (a rest frame) are

$$g_{ij} = \left(1 + \frac{M}{2r}\right)^4 \delta_{ij}; \qquad K_{ij} = 0. \tag{28}$$

We perform a Lorentz boost (on the isotropic x, y, z coordinates) of speed v in the (negative) z direction. We keep terms through $O(v^2)$ and $O(r^{-1})$ in g_{ij} and $O(v)$ and $O(r^{-2})$ in K_{ij} so as to obtain the lowest order nontrivial results. A straightforward calculation can employ either ordinary coordinate transformations in the Schwarzschild spacetime [with $\beta_i = 0$ and $\alpha = (1 - M/2r)(1 + M/2r)^{-1}$] or the gauge method using $\partial/\partial\xi$ with the ξs growing "like r" as they do for boosts. I used the former method with

$$x = \bar{x}, \qquad y = \bar{y}, \qquad z = \bar{z}(1 + \tfrac{1}{2}v^2) - v\bar{t},$$
$$t = \bar{t}(1 + \tfrac{1}{2}v^2) - v\bar{z} - \tfrac{1}{2}Mv(\bar{z}/\bar{r}). \tag{29}$$

The last term in the t-transformation is a gauge term chosen to make $\mathrm{tr}_{\bar{g}}\bar{K}_{ij} = O(\bar{r}^{-3})$ on the boosted slice $\bar{t} = 0$. I find $\bar{g}_{ij}$ and $\bar{K}_{ij}$ on the boosted slice in the coordinates $\bar{x}$, $\bar{y}$, $\bar{z}$ ($\bar{r}^2 = \bar{x}^2 + \bar{y}^2 + \bar{z}^2$). (This is the "push-forward" of g_{ij}, K_{ij} rather than the "pull-back.") From the boosted data one calculates, as expected,

$$\bar{E} = M + \tfrac{1}{2}Mv^2; \qquad \bar{P}^i = (0, 0, Mv). \tag{30}$$

The form (28) of the original metric suggests a definition of energy as essentially the $O(r^{-1})$ part of the conformal factor of the metric. To make this more precise, suppose we write, as is always possible,

$$g_{ij} = \psi^4 \delta_{ij} + \tilde{h}_{ij} \tag{31}$$

with $\psi = 1 + O(r^{-1})$ and $\tilde{h}_{ij} = O(r^{-1})$. Clearly we may define $\tilde{h}_{ij}$ and ψ such that $\mathrm{tr}_{\delta}\tilde{h}_{ij} = 0(r^{-2})$. Then, given g_{ij} and δ_{ij}, there is no ambiguity in the $O(r^{-1})$ part of ψ. Comparison with (28) suggests the definition [14,15]

$$E = -\frac{1}{2\pi} \oint_{\infty} (\nabla^i\psi)d^2S_i. \tag{32}$$

Another motivation for this formula is that it allows one to bring to bear conformal techniques on the positive energy conjecture. Comparing (32) and the gauge-invariant expression (20) shows that the following conditions are sufficient for the validity of (32):

$$g_{ij} = \psi^4\delta_{ij} + \tilde{h}_{ij}; \qquad \psi = 1 + O(r^{-1}),$$
$$\tilde{h}_{ij} = O(r^{-1}); \qquad \mathrm{tr}_{\delta}\tilde{h}_{ij} = O(r^{-2}), \tag{33}$$
$$\tilde{h}_{ij,j} = O(r^{-3}); \qquad K_{ij} = O(r^{-2}).$$

It should be emphasized that these are gauge-fixing conditions on the metric. They can always be achieved using a choice of spatial transformations as described in (8), granted the definition of asymptotically flat Cauchy data [(3)–(5)]. However, once Eqs. (33) are satisfied, one is restricted to transformations $\tilde{F} : x^i \to \bar{x}^i = \tilde{F}^i(x)$ such that $(\tilde{F}$–identity map$)$ = (Euclidean translations plus Euclidean rotations) + $O(r^{-1})$; thus,

$$\left[\frac{\partial \tilde{F}^i}{\partial x^j} - \delta^i_j \right] = A^i_j + O(r^{-2}). \tag{34}$$

There is no further *asymptotic* spatial coordinate freedom available. The "internal" coordinates are arbitrary, as always.

It has been assumed in studies of the positive energy conjecture [16] that one may take

$$\tilde{h}_{ij} = O(r^{-2}). \tag{35}$$

However, this requirement is simply too restrictive from a physical point of view. For, as I shall now describe, it cannot be satisfied by the data on a boosted slice in the asymptotic exterior Schwarzschild solution, which should certainly be counted as asymptotically flat Cauchy data.

Return now to the boosted Schwarzschild data and drop the bars. The transformations (29) readily yield a metric of the form (31) with

$$\tilde{h}_{ij} = O(r^{-1}); \quad |\tilde{h}_{ij}| \sim M v^2 r^{-1} = P^2 E^{-1} r^{-1} + O(r^{-2}), \tag{36}$$

$$K_{ij} = O(r^{-2}); \quad |K_{ij}| \sim M v r^{-2} = P r^{-2} + O(r^{-3}). \tag{37}$$

However, this $\tilde{h}_{ij}$ does not satisfy $\tilde{h}_{ij,j} = O(r^{-3})$. Therefore, as one may check, the $O(r^{-1})$ part of ψ is not $2E$ nor does the integral (32) produce the correct result $E = M + \frac{1}{2}Mv^2$. The $O(r^{-1})$ part of this ψ involves a function of the standard angle θ on a two-sphere; it is a "mass aspect." In order to achieve $\tilde{h}_{ij,j} = O(r^{-3})$, one appends to the x, y, z transformations in (29) a particular infinitesimal gauge transformation generator ξ^i such that $\xi^i_{,j} = O(r^{-1})$ and (Mv^2) is the infinitesimal parameter. Therefore, $g_{ij} \to g_{ij} + Mv^2 \times (\xi_{i,j} + \xi_{j,i})$, or

$$\tilde{h}_{ij} \to \tilde{h}_{ij} + Mv^2(\xi_{i,j} + \xi_{j,i} - \tfrac{2}{3}\delta_{ij}\xi_{k,k}), \tag{38}$$

$$\psi^4 \to \psi^4 + (Mv^2)\tfrac{2}{3}\xi_{k,k}. \tag{39}$$

From the form of $\tilde{h}_{ij}$ I have found the required ξ^i such that the modified $\tilde{h}_{ij}$ satisfies $\tilde{h}_{ij,j} = O(r^{-3})$. They are unique up to additive constants as expected. Then using this ξ^i in (39) automatically removes the "mass aspect" in ψ^4 and yields

$$g_{ij} = \left[1 + \frac{M + (1/2)Mv^2}{2r} \right]^4 \delta_{ij} + \tilde{h}_{ij} + O(r^{-2}). \tag{40}$$

However, $\tilde{h}_{ij}$ is still of $O(r^{-1})$. There do not exist *three* functions ξ^i that can remove the *five* independent $O(r^{-1})$ components of $\tilde{h}_{ij}$. However, the three divergence conditions, which are all one needs for the validity of (32), are easily satisfied.

The conclusion is that not *all* asymptotically flat Cauchy data with net nonvanishing linear momentum are conformally flat through $O(r^{-1})$ in the metric. [Incidentally, that this cannot be a coordinate gauge problem has been verified by (a) computing the three-dimensional conformal curvature tensor [17] of g_{ij}. Its dominant term [of $O(r^{-4})$] is invariant with respect to gauge transformations. (b) A gauge transformation of the time coordinate—introducing a slightly deformed slice—cannot affect the $O(r^{-1})$ part of the metric.] This is *not* to say that one cannot have $P^i \neq 0$ and $\tilde{h}_{ij} = O(r^{-2})$, however. In fact, there exist asymptotically flat and conformally flat vacuum solutions of the constraints with $P^i \neq 0$. Certain nontrivial examples have been constructed with no pathological properties [18,43]. Nevertheless, one is not, in general, justified in assuming $\tilde{h}_{ij} = O(r^{-2})$.

V. The Positive Energy Conjecture

The positive energy conjecture asserts, for regular asymptotically flat initial data on an $\mathbb{R}^3$ manifold, that (1) the ADM four-momentum is timelike ($|E| > |P^i|$), and (2) future-pointing ($E = P^0 > 0$), unless $P^\mu = 0$, which can occur only for Minkowski spacetime Cauchy data. Obviously, some conditions must be placed on the matter fields that are present. Also, it is clear that if one can prove $E > 0$ in *all* cases except for Minkowski data, where $E = 0$, then P^μ *must* be nonspacelike. [It is sometimes convenient to adopt "nonspacelike" in place of "timelike" in (2).] However, because of the nature of the attempts to prove the conjecture, I shall retain "future pointing" and "timelike" or "nonspacelike" as separate conditions, despite the consequent redundancy implied by the Lorentz covariance of P^μ. In this section I shall assume that *no matter fields are present* ($T_{\mu\nu} = 0$; "vacuum" constraints). In Section VI I shall consider the various energy conditions that are needed on the matter stress-energy tensor $T_{\mu\nu}$.

A number of physical arguments can be given to justify the extensive efforts that have gone into attempts to prove the truth of the positive energy conjecture. The basic point is that if the energy could be negative, it could be arbitrarily large and negative, from which a number of consequences for astrophysics, and the physics of particles and fields in space–time generally, would follow. Geroch [19] has discussed some of these in terms of energy extraction possibilities. Moreover, Jang and Wald [20] and

Jang [21] have shown interesting relations between the "cosmic censor-ship" hypothesis for singularities and the positive energy conjecture (see also Tipler *et al.* [22]). A recent review of the positive energy conjecture has also been given by Brill and Jang [44].

We first consider the case of "time-symmetric" initial data: $K_{ij} = 0$. The fundamental treatment of this problem, and the basic paper in the study of the positive energy conjecture, was that of Brill [14]. He assumed, besides $K_{ij} = 0$, that the metric is axisymmetric and can be written in the form $g_{ij} = \psi^4 \hat{g}_{ij}$, $\hat{g}_{ij} = f_{ij} + \tilde{h}_{ij}$, $\tilde{h}_{ij} = O(r^{-2})$. Time symmetry implies that P^μ is timelike and $\tilde{h}_{ij} = O(r^{-2})$ shows that the mass is the $O(r^{-1})$ part of ψ. The assumption on $\tilde{h}_{ij}$, while stronger than needed in the general case, is justified here because $P^i = 0$. Brill proved $E > 0$ by studying the equation for the conformal factor ψ. It is clear from his analysis that $E = 0$ only at flat space.

In a recent paper, Jang [23] has also studied the time-symmetric problem. He permits $\tilde{h}_{ij} = O(r^{-1})$ and allows singularities on the time-symmetric slice, provided any singularity that might be present is inside an apparent horizon [3], which in the present case is a minimal two-surface. He also assumes that in the past or future of the time-symmetric slice, there is a Cauchy slice $\mathbb{R}^3$ satisfying the requirements of the conjecture as stated above. Jang uses an expression for the energy derived from the work of Ashtekar and Hansen [24], which, in turn, is closely related to the formula of Geroch [19]

$$ E = (\text{const}) \cdot \oint_{\text{large } r} (2R - \chi^2) r \, d^2 S, \tag{41} $$

where $\mathscr{R}$ is the intrinsic scalar curvature of the two-surface S of integration and χ is the trace of the extrinsic curvature of S as embedded in the slice Σ. Following Geroch's idea, Jang shows that there exist surfaces on which the integral is nonnegative and that it must be nondecreasing as r increases. The possible existence of a minimal area two-surface in Σ complicates the analysis, but Jang carefully resolved the difficulties and obtained the positive energy theorem for spacetimes with time-symmetric slices.

The next increase in generality comes from dropping the $K_{ij} = 0$ require-ment and retaining only the existence of maximal slices, i.e., $K_{ij} \neq 0$ but $\text{tr}_g K = 0$. An important result in this case was obtained by Ō Murchadha and York [15] (see Cantor [5]): Every initial data set with $\text{tr}_g K = 0$ has energy greater than, or equal to, the energy of some time-symmetric initial data. In this theorem there are no restrictions that $\tilde{h}_{ij} = O(r^{-2})$, nor indeed are there any gauge conditions such as $\hat{h}_{ij,j} = O(r^{-3})$ imposed. [The $O(r^{-1})$ part of ψ in $\bar{g}_{ij} = \psi^4 g_{ij}$ captures the energy *difference* in $\bar{g}_{ij}$ and g_{ij}

in a gauge-invariant way.] The essential feature of this theorem, and that of Jang, is that if we have $\text{tr}_g K = 0$ and the Hamiltonian constraint (1) in vacuum, then $R \geq 0$. So any initial data set with nonnegative scalar curvature has nonnegative energy, in accordance with the conjecture.

However, the above conclusion does *not* resolve the full conjecture even for space–times that contain maximal slices. This is because, in general, such spaces *may* have $P^i \neq 0$. In such cases it would be necessary to show not only that P^μ is future-pointing ($E > 0$) but also that $|P^i| < E$ (timelike) or $|P^i| \leq E$ (nonspacelike). Now the assumption of a time-symmetric slice $K_{ij} = 0$ means $P^i = 0$ and there are no problems. However, if $K_{ij} \neq 0$, with $\text{tr}_g K = 0$, satisfies the *momentum constraint* (2) (with $j^i = 0$), which has thus far been ignored, $K_{ij} = O(r^{-2})$ is allowed and consequently we can have $P^i \neq 0$. One would have to show from the *combined* constraints that $|P^i| \leq E$ to get the full theorem for spacetimes with maximal slices. I am not aware that this has ever been done directly. On the other hand, there are no difficulties here. High-curvature, high-velocity vacuum initial data sets with $|P^i| \neq 0$ have been constructed [18] and in no instance has it been found that either $E > 0$ or $E > |P^i|$ was violated.

Despite these reservations, and the possibility that the maximal slice condition cannot always be satisfied, this theorem represents a great advance because it always permits $K_{ij} = O(r^{-3})$, which implies $P^i = 0$, but is *not* as strong as a time-symmetry requirement. In other words, there exist data sets with $\text{tr}_g K = 0$, $K_{ij} = O(r^{-3})$, whose developments have *no* time-symmetric slices. (For example, choose a "(t, ϕ)-symmetric" K_{ij} on the initial slice [25].) Clearly there also exist [18] tensors $K_{ij} = O(r^{-2})$ with $\text{tr}_g K = 0$ whose linear momenta will satisfy $|P^i| < E$. In these cases, also, P^μ is timelike.

We are now in a position to summarize the situation when $\text{tr}_g K = 0$ and no sources are present. Let $k^i_{(a)}$ denote the three-unit orthogonal translational Killing vectors of the flat metric to which g_{ij} is asymptotic. Then we have that the complete positive energy conjecture ($P^0 \geq 0$ and P^μ timelike except for Minkowski data, where $P^\mu = 0$) is true for asymptotically flat, source-free Cauchy data sets with

(a) $\text{tr}_g K = 0$,

(b) $\displaystyle\sum_{a=1}^{3}\left[\oint_\infty K^i_j k^j_{(a)}\, d^2 S_i\right]^2 < E^2.$

Condition (b) obviously includes $P^i = 0$. We may remark that the space of initial data that satisfy (a) and (b) is infinite-dimensional [4,5]. This theorem results from the ideas and work of a number of people: Brill, Geroch, Jang, Ō Murchadha, York, and, needless to say, Arnowitt, Deser, and Misner.

One may say of condition (b) that it is merely stating that P^μ is timelike and that this is not really helpful. However, it can indeed be helpful. For example, one can *always* determine, when $\mathrm{tr}_g K = 0$, that $P^i = 0$ (and thus $E > |P^i|$), *without* knowing the value of E. This follows immediately from the conformal method of solving the full constraint problem (see, e.g. [1]).

Schoen and Yau [16] have recently published a proof of the conjecture for the case when $\mathrm{tr}_g K = 0$. They allowed sources ($\mu \neq 0$, $j^i \neq 0$), but we shall defer consideration of sources until the next section.and take $\mu = j^i = 0$ here. Their method of proof was to show, by constructions using conformal techniques and some of their own important results in minimal surface theory, that $E < 0$ leads to a contradiction, and therefore that $E \geq 0$ in accordance with the conjecture. However, this work does not resolve the *complete* conjecture, even granted the maximal slice condition. First, as we have seen above, the issue of whether $|P^i| \leq E$ is fundamental, and this was not considered. Second, they assumed that one can always take the metric in the form $g_{ij} = [1 + (E/2r)]^4 \delta_{ij} + \tilde{h}_{ij}$, $\tilde{h}_{ij} = O(r^{-2})$. We have seen that this fast fall-off condition is not, in general, justified. I understand [26] that the Schoen–Yau construction [16] has difficulties that have not been resolved when the fall-off requirement is relaxed to $\tilde{h}_{ij} = O(r^{-1})$; $\mathrm{tr}\ \tilde{h} = O(r^{-2})$; $\tilde{h}_{ij,\,j} = O(r^{-3})$.

Now we turn to other approaches to prove the conjecture that, in effect, bypass having to demonstrate directly that $|E| \geq |P^i|$. I refer to the efforts of Brill and Deser [27] and Choquet-Bruhat and Marsden [28] to show that (a) Minkowski data comprise the only critical point of the energy expression (18) considered as a functional of the vacuum initial data that satisfy the constraints, and (b) the energy in a suitably defined neighborhood of Minkowski data is positive (the critical point is a local minimum). Actually, Brill and Deser seemed originally to believe that their results showed more than a *local* minimum. A mathematical criticism of their analysis has been given by Choquet-Bruhat and Marsden [28]. However, I will argue here that the Choquet-Bruhat and Marsden proof, while mathematically impeccable, does not actually prove the full conjecture *even near Minkowski data*, unless it is supplemented by an important and correct result obtained by Brill and Deser.

Both pairs of authors assumed the existence of maximal slices in a functional neighborhood of Minkowski data and we now know that this is fully justified [29,30]. A beautiful formulation by Choquet-Bruhat and Marsden showed, even *without* the maximal slice condition, that Minkowski data comprise the *only* critical point of the energy functional. However, in carrying out their second variations to show that the mass is positive, they did not vary the K_{ij}s because, as they noted, the $\overline{\mathrm{O}}$ Murchadha–York result implies that one need only show that time-symmetric data ($K_{ij} = 0$)

have positive mass, since we are assured that $\text{tr}_g\, K = 0$ in the present case. However, we have seen the danger to which this simplification leads. One may show that the energy is positive without showing that P^μ is timelike, or, at least, non spacelike.

However, the Choquet-Bruhat and Marsden analysis can be amended to satisfy this objection. All one has to do is to consider the variations of the K_{ij}s (in the second variation of the energy functional) that are compatible with the vacuum momentum constraints and $\text{tr}_g\, K = 0$. If these added variations lead to a positive contribution to the mass, and they *do*, as shown by Brill and Deser [27], then one knows $E \geq 0$ and P^μ is nonspacelike. For one already knows that Minkowski data are the only critical point. From Lorentz covariance of P^μ, therefore, $E > 0$ in *all* frames (except at the critical point), a condition that rules out spacelike P^μ. Further work, along the lines suggested by Deser [31], would rule out the case of null P^μ. Here one would expect only "plane-wave" type solutions that could not be asymptotically flat. The conclusion is that, near Minkowski data, the complete vacuum positive energy conjecture has been shown to be true, as follows from the work of Brill and Deser, the theorem of Choquet-Bruhat and Marsden, and the present critique.

Another powerful approach to the conjecture in source-free space–times follows from recent work of Jang [32]. This will be described in the next section.

It should also be noted that the positivity of energy, and the nonspacelike character of P^μ, have been demonstrated formally by considering classical gravity as an appropriate limit of supergravity [33–35]. The present status of this work seems to be that it is highly suggestive and intriguing, but does not constitute proof at the level of precision that most students of the problem find satisfactory.

VI. Conditions on the Stress-Energy Tensor

The positive energy conjecture in spacetimes with matter requires conditions on the allowable stress-energy tensors, as two simple examples below will illustrate. The three energy conditions most often discussed are the ones figuring in the proofs of singularity theorems. In order to express these in a convenient form, we define

n^μ as the timelike unit normal of slice Σ ($n_\mu n^\mu = -1$),

$\perp^\mu_\nu = \delta^\mu_\nu + n^\mu n_\nu$ as the operator of projection onto Σ,

$\mu = T^{\mu\nu} n_\mu n_\nu$ as the matter energy density on Σ,

$j^\mu = -\perp^\mu_\nu T^{\nu\alpha} n_\alpha$ as the matter momentum density on Σ,

$S^\mu_\nu = \perp^\mu_\alpha \perp^\beta_\nu T^\alpha_\beta$ as the matter stress tensor on Σ.

Then we have [3]

Weak energy condition: $\mu \geq 0$,
Strong energy condition: $\mu + \operatorname{tr} S \geq 0$,
Dominance of energy condition: $\mu \geq (j^\alpha j_\alpha)^{1/2}$.

Note that "dominance" implies "weak."

A simple example of Jang [36] shows the utility of the dominance condition. Let $g_{ij} = \delta_{ij}$ and $K_{ij} = \frac{1}{3} f \delta_{ij}$, where f is a smooth function of compact support. These data satisfy the constraints (1) and (2) with $\mu = (24\pi)^{-1} f^2$ and $j_i = -(12\pi)^{-1} f_{,i}$. The weak energy condition is satisfied, but $E = 0$, $P^i = 0$, and the data are not Minkowskian ($\mu \neq 0, j_i \neq 0, K_{ij} \neq 0$). However, the dominance condition $\mu \geq (j_i j^i)^{1/2}$ easily eliminates counter-examples of this type.

The next example shows that, even with $\operatorname{tr}_g K = 0$, $E > 0$, $\mu \geq 0$, one can find nonvacuum data arbitrarily close to Minkowski data for which P^μ is spacelike. Take $\mu = \varepsilon$, $\varepsilon =$ an arbitrarily small and positive constant, for $r \leq r_0$ and $\mu = 0$ for $r > r_0$. Take $j^i = \varepsilon V \delta^i_z$, with $V =$ constant > 1, for $r \leq r_0$ and $j^i = 0$ for $r > r_0$. Then the metric, for $r > r_0$, is $g_{ij} = \delta_{ij}[1 + \frac{1}{3}\varepsilon r_0^3 r^{-1}]$ $+ O(\varepsilon^2)$ and, for $r > r_0$, $K_{ij} = (\frac{3}{2}r^2)[P_i n_j + P_j n_i - (\delta_{ij} - n_i n_j)P^k n_k] +$ $O(\varepsilon^2) + O(r^{-4})$, where $n_i = x^i/r$ ($x^i = x, y, z$) and $P_i = \frac{1}{6}\varepsilon r_0^3 V \delta_{iz}$. I shall not write down the further unnecessary details of g_{ij} and K_{ij}. We see that $E = \frac{1}{6}\varepsilon r_0^3 + O(\varepsilon^2)$ and $|P^i| = EV$, $V > 1$. Therefore P^μ is spacelike.

These examples show clearly that the weak energy condition (or weak plus strong) is (are) insufficient to prove the conjecture. That dominance would probably be sufficient was pointed out by Geroch [37]. Recall that the basic techniques of proving the truth of the conjecture, when $\operatorname{tr}_g K = 0$, in the vacuum case relied only on $R(g) \geq 0$, which follows from the Hamiltonian constraint (1). If we assume the weak condition $\mu \leq 0$, we still have $R(g) \geq 0$ and the same techniques allow us to conclude $E \geq 0$. However, just as in the vacuum case, one has failed to analyze the momentum. In case $j^i \neq 0$, as here, this suggests the usefulness of a condition such as $\mu \geq (j_i j^i)^{1/2}$. Therefore, if the full conjecture could be proven true, with the dominance condition providing a handle on the momentum constraints, the vacuum problem would be completely solved. Dominance of energy would then be a sufficient condition on the matter fields for the positive energy theorem. However, it is unlikely that such a local condition would turn out also to be necessary for positive energy. (One could still have positive energy if the condition happened to be violated somewhere, but not everywhere.)

In another generalization of Geroch's approach, Jang [32] has effectively reduced the entire conjecture in the general case ($\operatorname{tr}_g K \neq 0, \mu \neq 0, j^i \neq 0$), assuming dominance, to the issue of whether or not a certain quasi-linear differential equation of elliptic type ("Jang's equation," known technically as an equation of the "mean curvature type") has a solution with satisfactory asymptotic growth. If it does, then Jang shows that an energy integral of the "Geroch type" (41) is always nondecreasing and the conclusion could be reached in the same manner described earlier in connection with Jang's analysis of the time-symmetric problem.

It appears then that Jang's equation may present the a problem in the positive energy analysis. For this reason I shall sketch one version of its content, following Cantor and York [38] and point out some of the mathematical problems. For other descriptions, see Jang [32] and Schoen and Yau [39].

Suppose $(g_{ij}, K_{ij}, \mu, j^i)$ are the initial data on Σ and that the constraints are satisfied, together with the dominance of energy condition. Suppose further that these data possess energy E and linear momentum $\mathbf{P}$. Introduce on Σ a scalar function W such that $|\nabla W| = O(r^{-1})$. Define an "auxiliary" metric and its inverse

$$\bar{g}_{ij} = g_{ij} + (\nabla_i W)(\nabla_j W), \tag{42}$$

$$\bar{g}^{ij} = g^{ij} - \frac{(\nabla^i W)(\nabla^j W)}{[1 + (\nabla W)^2]}. \tag{43}$$

Note that $E[g] = E[\bar{g}]$. (Indices on $\nabla_i W$ are raised and lowered with g_{ij}.) Also introduce an auxiliary second fundamental tensor

$$\bar{K}_{ij} = K_{ij} - \frac{\nabla_i \nabla_j W}{[1 + (\nabla W)^2]^{1/2}} \tag{44}$$

Geroch (see Jang [32]) has shown that transformations (42) and (44) are, if $\bar{g}_{ij}$ is flat and $\bar{K}_{ij}$ is zero, "gauge" transformations from a spacelike hyperplane to a spacelike curved surface in Minkowski spacetime.

Using the same techniques as in the discussion of the gauge invariance of $\mathbf{P}$, one can show that $\mathbf{P}[\bar{K}, \bar{g}] = \mathbf{P}[K, g]$. Jang's equation may be obtained by demanding a maximal slice condition holds for the modified data (which, however, no longer satisfy the constraints).

$$\mathrm{tr}_{\bar{g}} \bar{K} \equiv \bar{g}^{ij} \bar{K}_{ij} = 0, \tag{45}$$

or, in equivalent useful forms:

$$\left(g^{ij} - \frac{(\nabla^i W)(\nabla^j W)}{[1 + (\nabla W)^2]}\right)\left(K_{ij} - \frac{\nabla_i \nabla_j W}{[1 + (\nabla W)^2]^{1/2}}\right) = 0, \tag{46}$$

$$\frac{\Delta W}{[1 + (\nabla W)^2]^{1/2}} - \frac{(\nabla_i W)(\nabla_j W)(\nabla^i \nabla^j W)}{[1 + (\nabla W)^2]^{3/2}} = \mathrm{tr}_g K - \frac{K_{ij}(\nabla^i W)(\nabla^j W)}{[1 + (\nabla W)^2]}. \tag{47}$$

The second-order operator is the left-hand side of (47), which can also be written as a divergence,

$$\nabla_i \{(\nabla^i W)[1 + (\nabla W)^2]^{-1/2}\}, \tag{48}$$

or derived by varying W in

$$\int_\Sigma [1 + (\nabla W)^2]^{1/2} \, dv. \tag{49}$$

If we multiply the left side of (47) by $[1 + (\nabla W)^2]^{3/2}$ and construct the "symbol" $\sigma(\xi)$ of the resulting operator, we find

$$\sigma(\xi) = \xi^2 + \xi^2 u^2 - (\xi \cdot u)^2, \tag{50}$$

where ξ and u are vectors and the scalar product uses g_{ij}. Note that $\sigma(\xi) \geq 0$ and is zero only if $\xi = 0$, for arbitrary u. Therefore the operator is elliptic. From these conclusions, it can be stated that the only fundamental difficulty in resolving Jang's equation in the physically appropriate $M^p_{s, \delta}$ spaces is to demonstrate the boundedness of $|\nabla W|$.

Schoen and Yau [39] have given an outline of the resolution of Jang's equation. However, they assumed the overly restrictive fall-off condition on g_{ij}, described earlier.

VII.   Angular Momentum

The spatial angular momentum vector **J** is found from a two-surface integral exactly like the one for linear momentum, except that ϕ^i is now a "rotational Killing vector of the flat metric to which g_{ij} is asymptotic." An essential problem in showing that **J** is well defined for asymptotically flat Cauchy data, that is, that **J** is asymptotically gauge invariant, is that the phrase above enclosed by quotation marks is inherently ambiguous, unless ϕ^i is an exact symmetry (Killing vector) of g_{ij}. The ambiguity arises because there are, in effect, an infinite number of ways to write $g_{ij} =$ (flat three-metric) + h_{ij}, i.e., the coordinate gauge problem. There *is* no problem if a globally preferred structure like a Killing vector exists. Without exact symmetry, one cannot distinguish in any clear manner between the action of ϕ^i on the data and the action of an asymptotic gauge transformation on the data. (A similar problem is well known also in defining angular momentum at null infinity.) These problems do not arise for translations.

For example, suppose the data (g_{ij}, K_{ij}) possess linear momentum P^i. Then the asymptotic form of the part of K_{ij} containing P^i is [1]

$$K_{ij} = \frac{3}{2r^2} [P_i s_j + P_j s_i - (f_{ij} - s_i s_j) P^k s_k], \tag{51}$$

where f_{ij} is the flat metric and s^i is the unit normal of the sphere of integration. Then from (24) and (25) one can see that any ξ^i such that (a) $\xi^i_{,j} = O(r^{-1})$, and (b) ξ^i is not an odd function on the sphere, will lead to a change in $\mathbf{J}$. Of course, one expects "origin dependence" for *constant* translations of the origin, if $\mathbf{P} \neq 0$. However, here the situation is clouded by the generalized *spatial* translations, which are not constants. This ambiguity can be eliminated by gauge-fixing conditions such as $\tilde{h}_{ii} = O(r^{-2})$; $\tilde{h}_{ij,j} = O(r^{-3})$ [1]. These conditions eliminate everything except "internal" coordinate transformations, constant translations, and Euclidean rotations [with respect to which $\mathbf{J}$ transforms, via (24), as expected]. Therefore, the disease caused by generalized spatial translations can be cured by gauge conditions.

A detailed examination of (25) shows, however, that generalized time translations [$\xi^* = O(1)$] cause problems. *Even* if we eliminate these by requiring tr $K = O(r^{-3})$, the problem does not go away in general. Here we must require that either (a) ϕ^i is an *exact* rotational symmetry of g_{ij} [which is obvious from the form of (25)] or (b) the components of the Ricci tensor of g_{ij} defined by $R_{ij}s^i\phi^j|\phi|^{-1} = O(r^{-(3+\varepsilon)})$, $\varepsilon > 0$. The latter seems to be a physical restriction whose general significance is unknown to me at present, except to say that it permits the existence of "almost rotational symmetries" [41], which are precisely what one needs here. Of course, if (a) holds, then we may forget about gauge conditions and simply assert that the angular momentum is well defined.

One might think that $\mathbf{J}$ would at least be well defined in frames in which $\mathbf{P} = 0$. For them, the angular momentum is "intrinsic" in the same sense that it could, in analogy with flat spacetime theories, be independent of the location of an "origin," even in the generalized sense. I do not have a complete proof that this is always true; however, it is sufficient that the asymptotically dominant term of K_{ij} has the form [1]

$$K_{ij} = \frac{3}{r^3} [\varepsilon_{kil} J^l s^k s_j + \varepsilon_{kjl} J^l s^k s_i], \tag{52}$$

where J^l is the intrinsic angular momentum, and ε_{kil} is the Levi–Civita tensor.

Recently, angular momentum has been reformulated and studied by Ashtekar and Hansen [24] and by Sommers [42]. Their ideas could help clarify some of the issues raised above, together with the problem of defining a spacetime angular momentum tensor. See also Winicour [45].

Acknowledgments

I thank N. Ō Murchadha, L. Smarr, P. Sommers, F. J. Tipler, and J. A. Wheeler for helpful discussions. I am grateful to the Center for Theoretical Physics and the Center for Relativity at the University of Texas, Austin, for help and hospitality while this article was being written. The work was supported in part by a grant from the National Science Foundation.

References

[1]   York, J. W., Jr., *in* "Sources of Gravitational Radiation" (L. Smarr, ed.), pp. 83–126. Cambridge Univ. Press, London and New York, 1979.

[2]   Misner, C. W., Thorne, K. S., and Wheeler, J. A., "Gravitation." Freeman, San Francisco, California, 1973.

[3]   Hawking, S. W., and Ellis, G. F. R., "The Large Scale Structure of Spacetime." Cambridge Univ. Press, London and New York, 1973.

[4]   Cantor, M., *Commun. Math. Phys.* **57**, 83 (1977).

[5]   Cantor, M., *J. Math. Phys.* **20**, 1741 (1979).

[6]   Wheeler, J. A., *in* "Relativity, Groups, and Topology" (C. DeWitt and B. DeWitt, eds.), pp. 315–520. Gordon & Breach, New York, 1963.

[7]   Smarr, L., and York, J. W., Jr., *Phys. Rev. D* **17**, 2529 (1978).

[8]   Ó Murchadha, N., "Gravitational Energy and Momentum," Preprint, University College, Cork, Ireland, December, 1978.

[9]   Choquet-Bruhat, Y., and Geroch, R., *Commun. Math. Phys.* **14**, 329 (1969).

[10] Arnowitt, R., Deser, S., and Misner, C. W., *in* "Gravitation" (L. Witten, ed.), pp. 227–265. Wiley, New York, 1962.

[11] Misner, C. W., *Phys. Rev.* **130**, 1590 (1963).

[12] Geroch, R., *J. Math. Phys.* **13**, 956 (1972).

[13] York, J. W., Jr., "Has the Positive Energy Conjecture been Proved?" Preprint, University of North Carolina, Chapel Hill, December, 1978.

[14] Brill, D., *Ann. Phys. (N.Y.)* **7**, 466 (1959).

[15] Ó Murchadha, N., and York, J. W., Jr., *Phys. Rev. D* **10**, 2345 (1974).

[16] Schoen, R., and Yau, S.-T., *Commun. Math. Phys.* **65**, 45 (1979).

[17] York, J. W., Jr., *Phys. Rev. Lett.* **26**, 1659 (1971).

[18] York, J. W., Jr., and Piran, T., "The Initial Value Problem and Beyond," to appear in "The Alfred Schild Memorial Lectures" (R. Matzner and L. Shepley, eds.).

[19] Geroch, R., *Ann. N.Y. Acad. Sci.* **224**, 108 (1973).

[20] Jang, P. S., and Wald, R. M., *J. Math. Phys.* **18**, 41 (1977).

[21] Jang, P. S., *Phys. Rev. D* **20**, 834 (1979).

[22] Tipler, F. J., Clarke, C. J. S., and Ellis, G. F. R., *in* "General Relativity and Gravitation" (A. Held and P. Bergman, eds.), Vol. 2. Plenum, New York. In press.

[23] Jang, P. S., *Commun. Math. Phys.* To appear (1980).

[24] Ashtekar, A., and Hansen, R. D., *J. Math. Phys.* **19**, 1542 (1978).

[25] Hawking, S. W., *in* "Black Holes" (C. DeWitt and B. S. DeWitt, eds.), pp. 1–57. Gordon & Breach, New York, 1973.

[26] Schoen, R., personal communication, 1979.

[27] Brill, D., and Deser, S., *Ann. Phys. (N.Y.)* **50**, 548 (1968).

[28] Choquet-Bruhat, Y., and Marsden, J. E., *Commun. Math. Phys.* **51**, 283 (1976).

[29] Cantor, M., Fischer, A., Marsden, J., Ó Murchadha, N., and York, J., *Commun. Math. Phys.* **49**, 187 (1976).

[30] Choquet-Bruhat, Y., *C.R. Acad. Sci.* **280**, 169 (1975).

[31] Deser, S., *Nuovo Cimento B* **55B**, 593 (1968).

[32] Jang, P. S., *J. Math. Phys.* **19**, 1152 (1978).

[33] Deser, S., and Teitelboim, C., *Phys. Rev. Lett.* **39**, 249 (1977).

[34] Grisaru, M. T., *Phys. Lett. B* **73**, 207 (1978).

[35] Deser, S., *Phys. Rev. D* **19**, 3165 (1979).

[36] Jang, P. S., written communication to F. J. Tipler, March, 1978.

[37] Geroch, R., in a seminar at Princeton Univ., Princeton New Jersey, 1972.

[38] Cantor, M., and York, J. W., "Positive Energy and Positive Action." In preparation, 1980.

[39] Schoen, R., and Yau, S.-T., *Phys. Rev. Lett.* **43**, 1457 (1979).
[40] Schoen, R., and Yau, S.-T., "The Energy and the Linear Momentum of Space-times in General Relativity," Preprint, Institute for Advanced Study, Princeton, April, 1980.
[41] Ō Murchadha, N., and York, J. W., Jr., *Phys. Rev. D* **10**, 428 (1974).
[42] Sommers, P., *J. Math. Phys.* **19**, 549 (1978).
[43] Bowen, J., and York, J. W., Jr., "Time-Asymmetric Initial Data for Black Holes and Black Hole Collisions," *Phys. Rev. D.* In press.
[44] Brill, D. R., and Jang, P. S., *in* "General Relativity and Gravitation" (A. Held and P. Bergman, eds.), Vol. 1. Plenum, New York. In press.
[45] Winicour, J., *in* "General Relativity and Gravitation" (A. Held and P. Bergman, eds.), Vol. 2. Plenum, New York. In press.

Note added in proof: In a more recent analysis, Schoen and Yau [40] have adopted the appropriate asymptotic conditions (33). They state that their previous resolution [39] of Jang's equation still holds under these more general conditions. Then, rather than following the methods of Geroch [19] and Jang [23, 32], they employ conformal techniques from Ō Murchadha and York [15] and their previous work [16, 39] to show that the energy expression (32) is positive except in the case of Minkowski data, where it is zero. They assumed dominance of energy but made no restriction to maximal slices. Hence, the positive energy theorem has been proven by Schoen and Yau in appropriate physical and mathematical generality. Cantor and York have constructed an alternative proof [38].

5

The Beam and Stay of the Taub Universe

John Archibald Wheeler

Center for Theoretical Physics
Department of Physics
University of Texas at Austin
Austin, Texas

Abstract

The single mode of gravitational radiation that activates the otherwise empty Taub model universe has the longest possible wavelength that will fit into that universe and an amplitude just sufficient to curve the geometry up into closure, via its "effective" content of mass energy, both kinetic and potential. A parameter m' of time asymmetry in the Taub family of models allows one to adjust the ratio kinetic/potential at the phase of time asymmetry. A sufficiently extreme value of this parameter, $m' = 10^{12}$, gives a universe that will live as long as a typical Friedmann "dust-dominated" model ("stay") $= 60 \times 19^9$ years, but will have a volume-at-maximum expansion smaller by a factor of 4.8×10^{10}, or a ("beam") $= [(\pi/2)$ (volume at maximum)$]^{1/3}$ that is smaller than that of the Friedmann model by a factor of 3.64×10^3. If with a universe so small it is nevertheless possible to secure a stretch of time quite adequate for the development of life, it is not clear what is the point of a larger universe. Neither is it clear how the anthropic principle of Dicke and Carter is to come to terms with this "missed chance for economy."

Of all considerations that argue that the Universe is a giant machine indifferent to life or man, none is more impressive than the fantastic disparity in scale between the Universe and the Earth. Nothing has done more to blunt this argument than Dicke's analysis [1]. He emphasizes that the relevant comparison is not volume of space *versus* volume of the Earth or volume of man, but lifetime of the Universe compared to the time required for the development of life and observership. To put it another way [2]: Why should it take a universe with $\sim 10^{11}$ galaxies, and $\sim 10^{11}$ stars per galaxy, or altogether $\sim 10^{22}$ stars, to develop life and observership on one planet of one star of that universe? If that is the requirement, why not meet that requirement in a more economical way? Be moderate in that economy move. Cut the number of stars back, not from $\sim 10^{22}$ to 1, but only to $\sim 10^{11}$. That is enough to make a galaxy with its almost uncountable multitude of stars and that ought to be enough. It would, if space were what counted; but time is what counts. The reduction in the mass of the Universe by a factor $\sim 10^{11}$ not only cuts its dimensions by a factor $\sim 10^{11}$ but also reduces the calculated time from $\sim 10^{11}$ years to ~ 1 year. This is not long enough to permit even one star to develop, let alone life and mind. From this point of view, it is not at all obvious that the immensity of the Universe is at all out of keeping with the anthropic principle of Dicke and Carter [3]—going back to Berkeley and Parmenides—that it is meaningless to speak of a Universe in which the speaker is not located ("*Esse est percipi*," Berkeley [4]).

It is most difficult to know how to assess the anthropic principle. Nothing more will be said about it here. The point of attention will rather be this: What relation is there between the size of a closed model universe, with the topology of a three-sphere, and the duration or "stay" of that universe from start to finish? In particular, what is the relation between the two quantities in the case of the Taub universe [5]? This model universe is of exceptional theoretical interest and simplicity in this respect that it is curved up into closure neither by matter nor by electromagnetic radiation, but by gravitational radiation alone, and this of the longest wavelength that will fit into a closed universe [6]. One knows very well that it takes more fields than geometry to provide any realistic treatment of the Universe; but it is attractive to be able to consider a model calling on one field alone, and that pure geometry, for its complete discussion.

To define the question and issue more precisely it is desirable to translate the volume of the universe into a length,

$$\text{beam} = \left[\left(\frac{\pi}{2} \right) \left(\begin{array}{c} \text{volume at phase of} \\ \text{maximum expansion} \end{array} \right) \right]^{1/3}, \tag{1}$$

which can be compared directly with the relevant time (also expressed for simplicity in geometrical units, centimeters of light travel time),

$$\text{stay} = \left[\begin{array}{c} \text{maximum proper time available} \\ \text{to any world line from big bang} \\ \text{to big crunch} \end{array} \right]. \tag{2}$$

Why it is convenient to introduce the factor $(\pi/2)$ shows most clearly in the familiar example of the Friedmann model of an ideal three-sphere uniformly filled with a pressureless dust of matter. There is a relation between proper time t referred to any dust particle and radius a of the three-sphere expressed in parametric form by the formulas

$$a = (a_0/2)(1 - \cos \eta), \tag{3}$$

$$t = (a_0/2)(\eta - \sin \eta), \tag{4}$$

$$\eta \text{ from } 0 \text{ to } 2\pi. \tag{5}$$

The volume at the phase of maximum expansion is $2\pi^2 a_0^3$. The beam according to definition (1) is πa_0. This quantity has a simple meaning. It marks the greatest proper distance by which two dust particles can be separated at the phase of maximum expansion. Moreover, this beam, translated into a time, gives also the stay for this model. Only for this Friedmann "dust-filled" universe among all the models we have examined does this equality hold,

$$(\text{stay}) = (\text{beam})(\text{Friedmann, dust}); \tag{6}$$

for every other model taken up here we find

$$(\text{stay}) < (\text{beam})(\text{other models considered here}). \tag{7}$$

For a model universe—such as the Taub universe—that does not have three-sphere geometry but does have three-sphere topology, we still adhere to definition (1). Now, however, the beam does not represent the maximum obtainable separation between two points at the phase of maximum expansion, but a certain kind of average of the maximal separations between pairs of points at that instant. It is a happy circumstance for the usability of this definition even for the case of a highly contorted geometry that there exists a unique and well-defined spacelike slice of zero-extrinsic curvature and therefore of maximal volume [7].

Before turning to the Taub universe, it is appropriate to consider Friedmann models that differ from one another only in the equation of state of the medium that fills the space. With the connection between pressure and density of mass energy given by a proportionality factor γ independent of position and time,

$$p = \gamma p, \tag{8}$$

one knows from the formula

$$p \, dV = -dE \tag{9}$$

as applied to a three-sphere of radius a,

$$\gamma \rho d(a^3) = -d(\rho a^3), \tag{10}$$

that the density varies as $1/a^{3\gamma+3}$. Inserting this expression on the right-hand side of the familiar relation connecting density with the sum of extrinsic and extrinsic curvatures, one has the standard equation

$$a^{-2}(da/dt)^2 + a^{-2} = a_0^{3\gamma+1}/a^{3\gamma+3}.$$ (11)

Here the constant of proportionality in the formula for the density of mass energy has been expressed from the start in terms of the radius a_0 at maximum expansion. A simple integration gives

$$\text{(stay)} = \int dt = 2 \int_0^{a_0} (a_0^{3\gamma+1} a^{-3\gamma-1} - 1)^{-1/2} \, da$$

$$= \text{(beam)} \times \frac{[(1/2) - 3\gamma/(3\gamma+1)]!}{(1/2)![-3\gamma/(3\gamma+1)]!}.$$ (12)

A supplementary item of interest is the angular circuit η achieved during the entire stay by a photon, neutrino, or graviton that keeps going without deviation,

$$\eta = \int d\eta = \int dt/a(t) = 2 \int_0^{a_0} (a_0^{3\gamma+1} a^{-3\gamma-1} - 1)^{-1/2} \, da/a$$

$$= 2\pi/(3\gamma+1).$$ (13)

Both quantities appear in Table I.

It does not matter for the idealized model of the radiation-filled Friedmann universe (see Tolman [8], section on radiation dominated three-sphere model universe) ($\gamma = \frac{1}{3}$ in Table I) whether the radiation is composed of neutrinos or electromagnetic or gravitational radiation [6,9,10] or a mixture of the three so long as (1) the relevant wavelengths are all small compared to the scale of the universe, and (2) the distribution of radiation is isotropic in

Table I

Stay of Friedmann model universe from bang to crunch[a]

$\gamma = p/\rho$	0(dust)	$\frac{1}{3}$(radiation)	$\frac{2}{3}$	1("hard")
(stay)/(beam)	1.000	0.6366	0.4755	0.3814
η	2π	$2\pi/2$	$2\pi/3$	$2\pi/4$

[a] Expressed in terms of its beam for selected values of the parameter γ that measures the ratio of pressure to density of mass energy. Also listed is the angular circuit of the universe η achieved during the entire stay by a photon that keeps going without deviation.

direction and therefore homogeneous in position on each spacelike slice of constant t-value. The directly opposite case of a radiation-filled universe is that in which all of the radiant energy is concentrated in a single mode and this mode has the longest wavelength that will fit into the closed universe [6]. Then sphericity is lost. Then also it is expected to make a difference which kind of radiation energizes the system. In view of these complications, it is almost a miracle that Taub was able to find an exact solution [5] that is interpretable [6] as a universe deformed from sphericity by its content of effective gravitational wave energy.

To see this interpretation spelled out a little more fully, concentrate attention on the particular case of a Taub universe which is time symmetric in the sense that it attains maximum volume exactly halfway from "start" to "end." Focus on the three-geometry of this universe at the moment of time-symmetry,

$$ds^2 = l^2[4(d\psi + \cos\theta \, d\phi)^2 + (d\theta^2 + \sin^2\theta \, d\phi^2)]. \tag{14}$$

Compare this geometry with that of a three-sphere of radius a_0,

$$ds^2 = a_0^2[d\chi^2 + \sin^2\chi(d\alpha^2 + \sin^2\alpha \, d\beta^2)]. \tag{15}$$

Ask if a continuous transition can be brought about from (15) to (14) via a one-parameter family of intermediate geometries. For definiteness, demand that all the members of this family have the same volume $V = 2\pi^2 a_0^3$ or the same beam $(\pi V/2)^{1/3}$. Let them differ one from another solely in the fraction x of the effective mass energy that is in the form of homogeneous isotropic short-wave radiation:

(intrinsic curvature) + (extrinsic curvature)

$$= {}^{(3)}R + \begin{pmatrix} \text{zero at moment} \\ \text{of time symmetry} \end{pmatrix} = 16\pi \text{ (energy density)}$$

or

$$ {}^{(3)}R = 6x/a_0^2. \tag{16}$$

Here $x = 1$ corresponds to the case of the three-sphere (isotropic and therefore homogeneous short-wave radiation) and $x = 0$ corresponds to the case of the time-symmetric Taub universe (no "real" sources of mass energy; gravitational radiation only, and of maximal wavelength).

The appropriate mathematics is obtained by specializing the theory of the mixmaster model universe [11–13] (the relevant details of which can be summarized in three pages [14]) to a moment of time symmetry, and to deformations with one less parameter β_+ than the general mixmaster oscillation (β_+ and β_-). The metric components are expressed in a non-holonomic basis, one familiar from the study of the rotation group SO(3),

$$\sigma^1 = \cos \psi \, d\theta + \sin \psi \sin \theta \, d\phi,$$
$$\sigma^2 = \sin \psi \, d\theta - \cos \psi \sin \theta \, d\phi,$$
$$\sigma^3 = d\psi + \cos \theta \, d\phi. \tag{17}$$

"The simply connected covering space [for this group] has the three-sphere topology, and is obtained by extending the range of the Euler angle ψ to give it a 4π period" [14], so that

$$\int \sigma^1 \wedge \sigma^2 \wedge \sigma^3 = \int \sin \theta \, d\phi \wedge d\theta \wedge d\psi = 16\pi^2. \tag{18}$$

This figure is eight times as great as the integral over the unit sphere of (15),

$$\int \sin^2 \chi \, d\chi \sin \alpha \, d\alpha \, d\beta = 2\pi^2. \tag{19}$$

This factor eight is compensated by a halving of the distance factor that appears in each term of the space metric,

$$ds^2 = (a_0/2)^2 \{ e^{2\beta_+} [(\sigma^1)^2 + (\sigma^2)^2] + e^{-4\beta_+} (\sigma^3)^2 \}. \tag{20}$$

The volume remains $2\pi^2 a_0^3$ independent of the magnitude of the deformation parameter β_+. The scalar curvature invariant $^{(3)}R$ of the three-metric (20) is given in Eq. (13) of Misner et al. [14]. Inserting the expression for it into the equation of bending (16), and dividing out from both sides of the resulting equation the factor $(6/a_0^2)$, we arrive at a formula connecting (1) required deformation β_+ with (2) the fraction of the required energy density that is supplied by short-wave radiation,

$$x = (\tfrac{4}{3})e^{-2\beta_+} - (\tfrac{1}{3})e^{-8\beta_+}$$
$$= 1 - 8\beta_+^2 + (\tfrac{80}{3})\beta_+^3 - \cdots. \tag{21}$$

When short-wave radiation supplies the whole of the energy required to curve the model universe up into closure, then no deformation at all—that is, no longest-wave gravitational radiation—is needed; $x = 1$ implies $\beta_+ = 0$.

When a deformation is present, but when it is small enough so that we can neglect third and higher powers of the deformation in (21), then the relevant term is the one of second order. We see very clearly how in this limit the effective content of mass energy in the mode of gravitational radiation of longest wavelength is proportional to the square of the deformation from sphericity, as expected [6,9,10]. The constant of proportionality is not required here but may be recorded as of potential interest:

$$
\begin{pmatrix} \text{effective density of mass} \\ \text{energy of longest-wave} \\ \text{gravitational radiation} \end{pmatrix} \; (\text{in g/cm}^3)
$$

$$
= (1 - x) \times \begin{pmatrix} \text{density of 100\% short-wave} \\ \text{radiation required for closure} \\ \text{at phase of maximum expansion} \end{pmatrix}
$$

$$
= (1 - x)(3c^2/8\pi a_0^2 G) = 8\beta_+^2 (3c^2/8\pi a_0^2 G). \tag{22}
$$

The details of the alteration in metric required for this small $-\beta_+$ deformation can be read off from the zero- and first-order terms in the expansion of the ds^2 of (20) in powers of β^+; or, alternatively, from that one of the tensorial harmonics on a three-sphere, given by Lifshitz and Khalatnikov [11], which is of lowest index number and has the appropriate symmetry.

When no short-wave radiation at all is present, and the longest-wave radiation has to accomplish the whole feat of closure, then the scalar curvature $^{(3)}R$ at the phase of maximum expansion has to vanish altogether. Then $x = 0$ in Eq. (21). The deformation of the three-sphere has to have such a magnitude that

$$
e^{2\beta_+} = 4^{-1/3}; \qquad e^{-4\beta_+} = 4^{2/3}. \tag{23}
$$

Then the mixmaster metric (20) reads

$$
ds^2 = (a_0/2^{4/3})^2 [4(d\psi + \cos\theta \, d\phi)^2 + (d\theta^2 + \sin^2\theta \, d\phi^2)] \tag{24}
$$

This result agrees with the space part (14) of the time-symmetric Taub metric at the moment of time symmetry, provided that we identify the Taub l with

$$
l = a_0/2^{4/3}, \tag{25}
$$

as is also required by the condition that the volume be $2\pi^2 a_0^3$ or the beam be πa_0.

Thus for a geometry of zero-scalar curvature to curve up into the closure of a sphere would be impossible were the dimensionality two; the Gauss–Bonnet theorem

$$
\int {}^{(2)}R \, d(\text{surface}) = 8\pi(\text{two-sphere topology}) \tag{26}
$$

forbids. No such theorem applies in three dimensions. There the three-geometry can close not only when the scalar curvature invariant is zero, but even when it is negative. However, it is necessary that there be a positive principal curvature. In our case the three principal curvatures that contribute to $^{(3)}R$ are, respectively [14],

$$(2/a_0^2)[2 \exp(-2\beta_+ - 2\ 3^{1/2}\beta_-) - \exp(4\beta_+ + 4\ 3^{1/2}\beta_-)] = (7/2^{1/3}a_0^2),$$

$$(2/a_0^2)[2 \exp(-2\beta_+ + 2\ 3^{1/2}\beta_-) - \exp(4\beta_+ + 4\ 3^{1/2}\beta_-)] = (7/2^{1/3}a_0^2),$$

and

$$(2/a_0^2)[2 \exp(4\beta_+) - \exp(-8\beta_+)] = -(14/2^{1/3}\ a_0^2), \tag{27}$$

as compared to the value $2/a_0^2$ that each of these separate contributions has in a true three-sphere of the same volume. No one who has examined how the alternate focusing and defocusing of particle orbits in an alternating gradient cyclotron add up to a net focusing will be surprised that the alternating favorable and unfavorable geodesic deviations of geodesics in a Taub universe can add up, and do add up, to the net convergence typical of a closed-space geometry.

When we turn to the stay of the Taub universe, we have to recognize that this model universe, in its ideal version, contains no real big bang or big crunch, but only what Ellis and King [15] have called "whimpers" at those two times. However, Misner and Taub [16] have given reasons to believe that this ideal version is a set of measure zero among all nearby model universes, in the sense that any small perturbation, no matter how small, will grow to a singularity at both gates of time. Regardless of the ultimate truth of this conjecture [17–19], we have it in mind when we define the stay of the Taub universe as the longest interval of proper time that any geodesic can run between those two moments when the calculated volume of the ideal Taub universe assumes the value zero.

The stay of the Taub universe depends not only on its beam but also on the phase of the deformation oscillation at the instant of maximum expansion. The special case considered up to now has this deformation at its maximum in coincidence with the maximum in the volume. In another way of speaking, the effective energy of the gravitational wave was all potential, none kinetic, whereas we now consider cases where at maximum expansion this energy is part kinetic, part potential. The metric in this more general case is given by the expression [20]

$$ds^2 = -U^{-1}\ dt^2 + 4l^2U(d\psi + \cos\theta\ d\phi)^2$$
$$+ (t^2 + l^2)(d\theta^2 + \sin^2\theta\ d\phi)^2. \tag{28}$$

Here U is an abbreviation for the expression

$$U = [l^2 + m^2 - (t - m)^2]/(t^2 + l^2). \tag{29}$$

The constant m, or the dimensionless quantity

$$m' = m/l,$$ (30)

has to do with the phase of the gravitational wave at maximum expansion. The volume, the quantity that for a three-sphere would have the value $2\pi^2a^3$, has the magnitude

$$V = 32\pi^2 l^3 (1 + t'^2)^{1/2}(1 + 2m't' - t'^2)^{1/2},$$ (31)

where

$$t' = t/l.$$ (32)

For any given value of m' one can find at what value of the time parameter t' (not proper time!), the volume (31) assumes its maximum value. This problem leads to a cubic equation for t'. To solve it, write

$$m' \equiv m/l = 1/\sinh 3\sigma.$$ (33)

Then at maximum

$$t' \equiv t/l = 1/2 \sinh \sigma.$$ (34)

Calculating the volume at this time, multiplying it by $\pi/2$, and taking the cube root, we find the beam,

$$(\text{beam}) \equiv (\pi V_{\max}/2)^{1/3} = 2^{4/3}\pi l \frac{[1 + (1/4 \sinh^2 \sigma)]^{1/2}}{[1 + (3/4 \sinh^2 \sigma)]^{1/6}}.$$ (35)

The stay is

$$(\text{stay}) = \tau = \int d\tau = \int (t^2 + l^2)^{1/2}[l^2 + m^2 - (t - m)^2]^{-1/2} \, dt,$$ (36)

where the integration runs from one vanishing of the volume

$$t' = m' - (1 + m'^2)^{1/2},$$ (37)

to the other,

$$t' = m' + (1 + m'^2)^{1/2}.$$ (38)

It is more convenient to have the integration run from -1 to $+1$. This goal is achieved by the substitutions

$$m' \equiv m/l = \tan 2\lambda,$$ (39)

$$t' \equiv t/l = (z + \tan \lambda)/(1 - z \tan \lambda),$$ (40)

and gives

$(\text{stay}) =$

$$\int d\tau = l(1 - \tan^2 \lambda)^{1/2} \sec \lambda \int_{-1}^{1} [(1 + z^2)/(1 - z^2)]^{1/2}(1 - z \tan \lambda)^{-2} \, dz.$$ (41)

Table II

Stay and beam of the Taub universe for selected values of the parameter m' of time asymmetry.

$|m'|$	small	0.5	2	10	large
(stay)/l	$3.820^a + 1.554^b(m')^2$	4.187	7.64	21.5	$3.142^c|m'|$
(beam)/l	$7.916^d + 1.571^e(m')^{4/3}$	8.886	14.65	40.2	$8.64^f|m'|^{2/3}$
(stay)/(beam)	0.4826	0.4712	0.522	0.534	$0.364^g|m'|^{1/3}$

^{a} "3.820" $= 2^{3/2}E(k)$. Here E and K are the complete elliptic integrals of modulus $k = 2^{-1/2}$.
^{b} "1.554" $= 2^{-1/2}[3E(k) - K(k)]$.
^{c} "3.142" $= \pi$.
^{d} "7.916" $= 2^{4/3}\pi$.
^{e} "1.571" $= \pi/2$.
^{f} "8.64" $= 2^{2/3}3^{1/2}\pi$.
^{g} "0.364" $= 2^{-2/3}3^{-1/2}$.

From (35) (analytical) and (41) (numerical integration), we obtain the results for stay and beam listed in Table II.

For values of the parameter m' of asymmetry in time all the way from $m' = 0$ to $m' = 10$, the calculated ratio of stay to beam for the Taub universe, ranged only from ~ 0.47 to ~ 0.53. Thus it remains well below the value (stay)/(beam) $= 0.6366$ for a spherical model universe activated by short-wave radiation. In this respect none of these Taub universes is as "efficient in allowing time for life to develop" as is the dust-filled Friedmann model universe, with (stay)/(beam) $= 1.000$. However, when the parameter of time asymmetry is greatly increased, then the "efficiency" of the Taub universe surpasses that of every other model considered here, and increases without limit. As an example, consider the case $m' = 10^{12}$. Then the ratio of (stay)/(beam) is approximately 3.64×10^3. For such a model universe to live 60×10^9 yr, it need only have a beam of 16.48×10^6 lyr. In contrast, the usual "dust-dominated" Friedmann model has to have a beam of 60×10^9 lyr (equivalent to a radius-at-maximum of 19.1×10^9 lyr). The Taub universe is smaller in effective linear dimensions than the Friedmann universe by a factor of 3.64×10^3, and smaller in volume by a factor of 4.8×10^{10}. It does not need any matter at all to curve it up into closure. However, if this model universe contained matter roughly to the same density as the actual universe, that matter would suffice to make only of the order of a single galaxy.

If with a universe so small it is nevertheless possible to secure a stretch of time quite adequate for the development of life, what is the point of a larger universe? It is not obvious that the anthropic principle has any answer to this question.

Appreciation is expressed to Paul Gleichauf for a discussion.

Addendum on Lethal Radiation

The very large curvatures present in the earliest days of the Universe will inevitably lead to the production of primordial photons. The subsequent stretching of the geometry in one direction and squeezing in the perpendicular direction in the Taub universe (or in the more general mixmaster universe) will degrade photons traveling in the one direction but continually raise up the temperature of photons traveling in the other direction, as John Barrow notes in a much appreciated letter. He points out that enough such ultraviolet and x-ray radiation will rule out life as we know it. To tell what restrictions these considerations place on a "small-beam-long-stay" universe is an interesting question requiring a detailed investigation which is not undertaken here.

Appendix

We neglect l in comparison with m' and have for the volume

$$V = 32\pi^2 l^3 t'(2m't' - t'^2)^{1/2} \tag{42}$$

and for the proper time

$$\tau = l \int (2m't' - t'^2)^{-1/2} t' \, dt'; \tag{43}$$

or, with the parametrization

$$t' = m'(1 - \cos f)(f \text{ from } 0 \text{ to } \pi) \tag{44}$$

the results

$$\tau = lm'(f - \sin f), \tag{45}$$

$$(\text{stay})/l = \pi m', \tag{46}$$

and

$$V = 32\pi^2 l^3 (m')^2 \sin f(1 - \cos f) \tag{47}$$

which, assuming its maximum value at $\cos f = -1/2$, gives

$$(\text{beam})/l = 2^{2/3} \, 3^{1/2} \, \pi |m'|^{2/3}. \tag{48}$$

References

[1]   Dicke, R. H., *Nature (London)* **192**, 440–441 (1962).
[2]   Wheeler, J. A., "General Relativity and Gravitation (GR7)" (G. Shaviv and J. Rosen, eds.), pp. 299–344. Wiley, New York, 1975.
[3]   Carter, B., *in* "Confrontation of Cosmological Theories with Observational Data" (M. S. Longair, ed.), I.A.U., pp. 291–298. Reidel Publ., Dordrecht, Netherlands, 1974.
[4]   Berkeley, B., "Works" (A. C. Fraser, ed.), 3 vols. Clarendon, Oxford, 1871.
[5]   Taub, A. H., *Ann. Math.* **53**, 472–490 (1951).
[6]   Wheeler, J. A., *in* "Relativity, Groups and Topology" (C. DeWitt and B. S. DeWitt, eds.), pp. 242–307. Gordon & Breach, New York, 1964.
[7]   Marsden, J. E., and Tipler, F. J., "Maximal Hypersurfaces and Foliations of Constant Mean Curvature in General Relativity," preprint jointly from University of California at Berkeley and University of Texas at Austin, 1 May 1980.
[8]   Tolman, R. C., "Relativity, Thermodynamics, and Cosmology." Clarendon, Oxford, 1934.
[9]   Brill, D. R., and Hartle, J. B., *Phys. Rev. B* **135**, 271–278 (1964).
[10] Isaacson, R. A., *Phys. Rev.* **166**, 1272–1280 (1968).
[11] Lifshitz, E. M., and Khalatnikov, I. M., *Adv. Phys.* **12**, 185–249 (1963).
[12] Misner, C. W., *Phys. Rev. Lett.* **22**, 1071–1074 (1969).
[13] Belinsky, V. A., Khalatnikov, I. M., and Lifshitz, E. M., *Usp. Fiz. Nauk* **102**, 463–500 (1970); Engl. transl. *Adv. Phys.* **19**, 525–573 (1970).
[14] Misner, C. W., Thorne, K. S., and Wheeler, J. A., "Gravitation," pp. 806–808. Freeman, San Francisco, California, 1973.
[15] Ellis, G. F. R., and King, A. R., *Commun. Math. Phys.* **38**, 119–156 (1974).
[16] Misner, C. W., and Taub, A. H., *Zh. Eksp. Teor. Fiz.* **55**, 233–255 (1968); Engl. orig. *Sov. Phys.—JETP* **28**, 122–133 (1969).
[17] Belinsky, V. A., Khalatnikov, I. M., and Collins, B., "Dealing with the Instability of Special Solutions of the Field Equations," discussion paper, unpublished, June 1976.
[18] Wheeler, J. A., *Gen. Relativ. Grav.* **8**, 713–715 (1977).
[19] Barrow, J. D., and Matzner, R. A., *Mon. Not. Roy. Astr. Soc.* **181**, 719–728 (1977).
[20] Hawking, S. W., and Ellis, G. F. R., "The Large Scale Structure of Space-Time," pp. 170–171. Cambridge Univ. Press, London and New York, 1973.

6

Tidal Forces in a Highly Asymmetric Taub Universe[†]

L. C. Shepley

Department of Physics
The University of Texas at Austin
Austin, Texas

Abstract

Wheeler has given a highly asymmetric model universe that is a specific case of Taub space chosen to exhibit both a long lifetime and a small spatial volume. Here I show that the high asymmetry causes tidal forces to act on forming stars, but these tidal forces are too weak to disrupt the condensation process. Since stars can form, presumably life can, also. This example—so very unlike the real Universe—therefore is a difficulty for the anthropic principle.

I. Introduction

Why is the real Universe apparently unlike one of the very fascinating, unusual, and anisotropic models that have recently been studied by theoreticians? One answer is the conjecture that only an isotropic universe will allow the development of intelligent life; this anthropic principle suggests that an anisotropic universe would contain no observers. Yet the isotropic models are in a way inefficient. They achieve their long "stay" by having far more matter and far larger spatial size than is necessary for the support of life.

Wheeler [1] has recently investigated a limiting form of the Taub [2] metric and has shown that it has a remarkable and unexpected property. The total "stay" of this model universe between beginning bang and ending

† Supported by National Science Foundation grant PHY77-07619.

crunch can be made as long as one pleases, while yet the spatial size of the universe can be made as small as one pleases. When the value of Wheeler's parameter of time asymmetry m (defined below) is as great as 10^{12}, the total proper time can be made to have a value of cosmological magnitude, 6×10^{10} years. The volume of Wheeler's model, in the same example, can be made small enough to include enough matter for but one galaxy.

Wheeler's example seems to obey the requirement of the anthropic principle of Dicke [3] and Carter [4], to provide time enough to develop intelligent life (or at least as much time as a realistic isotropic model does), but does so much more economically. Is it not a difficulty for the anthropic principle, asks Wheeler, that nature does not take advantage of this possibility?

Could it be that Wheeler's model universe disobeys the anthropic principle after all? Does the great curvature anisotropy needed for long "stay" produce tidal forces so great that they would tear apart clouds of matter and prevent the formation of even a single sun? (Question asked by J. A. Wheeler in telephone conversation of 12 July 1978.)

This note answers "no": The tidal forces at the surface of a forming star in such a model universe are relatively 300 times stronger than the tidal effects of the moon on the oceans of Earth, but are not great enough to prevent star formation. Thus the question remains unanswered by the anthropic principle why nature employs our extravagant universe when a smaller one would do.

II. The Beam and Stay of the Taub Metric

Details of the analysis start with the metric of Taub space (see Misner and Taub [5]):

$$ds^2 = l^2\{-F^{-1}\,dt^2 + (1 + t^2)(\omega_1^2 + \omega_2^2) + 4F\omega_3^2\},$$

where

$$F = F(t) = (1 + 2mt - t^2)/(1 + t^2); \qquad m, l \text{ are constants};$$

and where ω_1, ω_2, and ω_3 are the one-forms:

$$\omega_1 = \cos\psi\,d\theta + \sin\psi\sin\theta\,d\phi,$$

$$\omega_2 = \sin\psi\,d\theta - \cos\psi\sin\theta\,d\phi,$$

$$\omega_3 = d\psi + \cos\theta\,d\phi.$$

The spatial coordinates are coordinates on the three-sphere and have the ranges $0 \le \psi \le 4\pi, 0 \le \phi \le 2\pi, 0 \le \theta \le \pi$. The constant m is a measure of time asymmetry in that the time from "bang" to maximum expansion equals the time from maximum to "crunch" only when $m = 0$ (m may be positive, negative, or zero, but is here assumed positive for convenience).

The coordinate times when F vanishes are

$$t_- = m - (m^2 + 1)^{1/2} \quad \text{and} \quad t_+ = m + (m^2 + 1)^{1/2}.$$

At these times the coordinate system breaks down, but there is no singularity in the Taub model, which is a model empty of all but gravitational energy. When even the slightest amount of ordinary matter is introduced, however, these apparent singularities become true singularities (Ryan and Shepley [6]). As a cosmological model, therefore, the proper lifetime or span S is

$$S = l \int_{t_-}^{t_+} F^{-1/2} \, dt = l \int_{t_-}^{t_+} \left[\frac{1 + t^2}{1 + 2mt - t^2} \right]^{1/2} dt.$$

With the change of variable $t = m + (1 + m^2)^{1/2} \sin x$, this integral becomes

$$S = \pi l (1 + m^2)^{1/2} I(m),$$

where

$$I(m) = \frac{1}{\pi} \int_{-\pi/2}^{\pi/2} [1 + 2m(1 + m^2)^{-1/2} \sin x + \sin^2 x]^{1/2} \, dx.$$

This form for S is convenient since

$$I(m) \approx 1;$$

the minimum value is $I(m) = 1$ when $m \to \pm \infty$, and the maximum value is $I(m) = 1.216$ when $m = 0$ (I wish to thank M. Rotenberry for help with this calculation).

The typical linear extent of the model $B(t)$ is defined as the circumference of the three-sphere that has the same volume as the space section of the model at a given time t. Thus

$$B(t) = 2\pi \left[\frac{V}{16\pi^2} \right]^{1/3} = 2^{4/3} \pi l [(1 + 2mt - t^2)(1 + t^2)]^{1/6}.$$

Wheeler calls the "beam" of the model the maximum value taken by $B(t)$.

In the limiting case of large positive m, the above analysis is greatly simplified. The metric becomes

$$ds^2 = l^2 \left\{ -\left(\frac{t}{2m - t} \right) dt^2 + t^2(\omega_1^2 + \omega_2^2) + 4 \left(\frac{2m - t}{t} \right) \omega_3^2 \right\}.$$

The span S is

$$S = l \int_0^{2m} \left[\frac{t}{2m - t} \right]^{1/2} dt = \pi l m,$$

and the linear extent is

$$B(t) = 2^{4/3} l\pi [t^3 (2m - t)]^{1/6}.$$

The coordinate time of maximum expansion (maximum B) is t_m, given by

$$t_m = \tfrac{3}{2}m,$$

at which time the maximum beam is

$$B_m = 2^{2/3} \, 3^{1/2} l\pi m^{2/3}.$$

The anthropic principle, at least, requires that the stay S be about the same as the lifetime of the closed Friedmann model that fits the real Universe well, namely, 6×10^{10} years, and that B_m be about the distance to a neighboring galaxy, say 2×10^7 years. (I am taking $c = G = 1$ so that one year is the basic unit of length, time, and mass. The acceleration of gravity at the surface of Earth is $10 \text{ m sec}^{-2} = 1 \text{ year}^{-1}$. The number density of stars in our galaxies is about one in a sphere of radius one year. The mass of a typical star like our own is about 2×10^{-13} year.) Hence the parameters l and m are

$$l = 2 \times 10^{-2} \text{ year},$$

$$m = 10^{12}.$$

III. Cosmic Tides

The effect of the anisotropic expansion of the model is to tend to tear apart localized condensations of matter. Let U be the four-velocity of the center of a condensing cloud, and let W be a vector pointing toward a point at the edge of the cloud. The cosmic tidal acceleration appears on the right side of the equation of geodesic deviation

$$W^\mu_{;\sigma\tau} U^\sigma U^\tau = R^\mu_{\sigma\tau\rho} U^\sigma U^\tau W^\rho,$$

and this acceleration must be overcome by the self-gravity of the condensing star if the star is to form. The components of the tidal acceleration are most easily computed in the orthonormal basis

$$\sigma_0 = lF^{-1/2} \, dt,$$

$$\sigma_1 = l(1 + t^2)^{1/2} \omega_1,$$

$$\sigma_2 = l(1 + t^2)^{1/2} \omega_2,$$

$$\sigma_3 = 2lF^{1/2} \omega_3.$$

Consider three particles at the surface of the cloud, each at proper distance r from the center, located along spatial directions i ($i = 1, 2, 3$). The cosmic tidal force on these particles produces the acceleration $a^\mu_{(i)}$:

$$a^\mu_{(i)} = rR^\mu_{00i},$$

where R^μ_{00i} is a component of the Riemann tensor in the orthonormal basis. The only nonzero curvature components of this form are

$$R^1_{001} = R^2_{002} = l^{-2}\left[(1 + t^2)^{-2}F + \tfrac{1}{2}t(1 + t^2)^{-1}\frac{dF}{dt}\right],$$

$$R^3_{003} = \tfrac{1}{2}l^{-2}\frac{d^2F}{dt^2},$$

or

$$R^1_{001} = R^2_{002} = -\tfrac{1}{2}R^3_{003} = \frac{1 + 3mt - 3t^2 - mt^3}{l^2(1 + t^2)^3}.$$

Incidentally, the other nonvanishing components of the Riemann tensor are

$$R^1_{203} = 2R^1_{302} = -2R^2_{301} = \frac{2(3t + mt^2 - m - t^3)}{l^2(1 + t^2)^3},$$

$$R^1_{313} = R^2_{323} = -\tfrac{1}{2}R^1_{212} = R^0_{101},$$

or else are found by use of the above results and the symmetries of the Riemann tensor.

In the large m approximation, these components are

$$R^2_{001} = R^2_{002} = -ml^{-2}t^{-3},$$

$$R^3_{003} = 2ml^{-2}t^{-3},$$

and indeed all nonvanishing components of the Riemann tensor are of magnitude $ml^{-2}t^{-3}$ or less.

Imagine that a star begins to form from cosmic dust at proper time $\tau = 3 \times 10^8$ years after the initial bang. The approximate relation of coordinate time t to proper time τ in this early situation is simply

$$\tau = \tfrac{1}{3}2^{1/2}lm^{-1/2}t^{3/2} = 10^{-8}t^{3/2}$$

Thus the proper time $\tau = 3 \times 10^8$ years corresponds to $t = 10^{11}$. A typical magnitude of tidal acceleration is $5 \times 10^{-18}r$ (in units of year^{-1}).

In any expanding model there is the problem of how a galaxy forms. Let us suppose a galaxy has formed in this model, however: It is a dust cloud of the average density of our own galaxy. (The linear extent at this time has already reached $B = 5 \times 10^6$ years so there is surely room enough for one galaxy.)

Suppose a star has started to form, a condensation of about one $M_\odot = 2 \times 10^{-13}$ year within a radius of $r =$ one year. The tidal accelerations at the surface of this cloud are about 5×10^{-18} year^{-1}. This mass exerts a gravitational acceleration of 2×10^{-13} year^{-1} at that distance, however, more than enough to overcome the tidal effects of the cosmos. (The tidal force of the moon at the surface of the Earth is about 10^{-7} of the Earth's gravity. Here the cosmic tidal force is 3×10^{-5} times the gravity due to the condensation.)

The tidal effects depend on t^{-3}. The dependence on proper time τ since the "bang" is as τ^{-2}. Presumably it takes about 10^8 years, at least, for a galaxy to form, so stars could form after the galaxy does in this model. At earlier times, however, cosmic anisotropy forces should keep things stirred up on the local level.

IV. Conclusion

A comparison with the isotropic case is useful. In that case all three terms R^i_{00i} $(i = 1, 2, 3)$ are equal; each is approximately ρ, where ρ is the cosmic density in geometrical units. At 3×10^8 years after the beginning bang, ρ has dropped to about 5×10^{-27} gm/cm^3 or $\rho = 4 \times 10^{-19}/$year^2. Each term rR^i_{00i} thus is smaller than in the present calculation. Also, in the isotropic case, once the condensing material has divorced itself from the background metric, the cosmos is free to expand and the star to contract without further mutual effect (Einstein and Straus [7]). This divorce procedure is far from trivial, of course; it is the largest problem in the theory of galaxy formation.

In the isotropic case, the lack of effect of cosmos upon star is shown by investigation of the metric within an empty spherical hole in the model. The Taub model, the present case, is itself devoid of all except gravitational content. The tidal effects are thus directly calculable. The cosmic tides pull and they squeeze; even though the cosmic anisotropy as measured by the asymmetry parameter m is great, they are too weak to disrupt the formation of stars. Notice that the cosmic effect does increase at times nearer to the "bang" than the 10^8 years of proper time assumed above. It may be that the tidal forces are sufficiently strong to prevent earlier formation of the massive stars needed for heavy element synthesis. In any case, J. Barrow (private communication to J. A. Wheeler) has raised the very interesting objection that the model may have other problems associated with the existence of blue-shifted photons.

It may be that details of astrophysics would prevent humanity from evolving in Wheeler's example. However, on the basis of the present calculation, apparently such evolution could take place: There is time enough;

there is room enough; and stars and planets are free to form. The anthropic principle thus seems to allow a universe much smaller than ours. Why then is ours so large and so isotropic?

References

[1] Wheeler, J. A., Chapter 5, this volume.
[2] Taub, A. H., *Ann. Math.* **53**, 472 (1951).
[3] Dicke, R. H., *Nature (London)* **192**, 440 (1962).
[4] Carter, B., *in* "Conformation of Cosmological Theories with Observational Data" (M. S. Longair, ed.), I. A. U., p. 291, Reidel Publ., Dordrecht, Netherlands, 1974.
[5] Misner, C. W., and Taub, A. H., *Zh. Eksp. Teor. Fiz.* **55**, 233 (1968); Engl. trans. *Sov. Phys. JETP* **28**, 122 (1969).
[6] Ryan, M. P., and Shepley, L. C., "Homogeneous Relativistic Cosmologies." Princeton Univ. Press, Princeton, New Jersey, 1975.
[7] Einstein, A., and Straus, E. G., *Rev. Mod. Phys.* **18**, 148 (1946); see also *Rev. Mod. Phys.* **17**, 120 (1945).

7

Symmetry Breaking in General Relativity[†]

Arthur E. Fischer

Department of Mathematics
University of California
Santa Cruz, California

Jerrold E. Marsden

Department of Mathematics
University of California
Berkeley, California

and

Vincent Moncrief

Department of Physics
Yale University
New Haven, Connecticut

Abstract

Bifurcation theory is used to analyze the space of solutions of Einstein's equations near a spacetime with symmetries. The methods developed here allow one to describe precisely how the symmetry is broken as one branches from a highly symmetric spacetime to nearby spacetimes with fewer symmetries, and finally to a generic solution with no symmetries. This phenomenon of symmetry breaking is associated with the fact that near symmetric solutions the space of solutions of Einstein's equations does not form a smooth manifold but rather has a conical structure. The geometric picture associated with this conical structure enables one to understand the breaking of symmetries. Although the results are described for pure gravity, they may be extended to classes of fields coupled to gravity, such as gauge theories. Since most of the known solutions of Einstein's equations have Killing symmetries, the study of how these symmetries are broken by small perturbations takes on considerable theoretical significance.

† Research for this work was supported in part by NSF grant PHY78-82353.

Bifurcation theory deals with the branching of solutions of non-linear equations. Here we describe a new application of this theory to the determination of how the solution set of Einstein's equations branches near a spacetime with a one-parameter family of symmetries. The directions of these branches are not determined by the linearized theory of gravity alone, but are completely characterized by the second-order terms. Thus the linearized theory of gravity near a spacetime with symmetry is not sufficient to capture the dominant effects of the nonlinear theory.

These conclusions are in accord with the earlier work of Fischer and Marsden [1–3], Moncrief [4–6], and Arms and Marsden [7], which showed that for spacetimes with compact Cauchy surfaces, a solution of Einstein's equations is linearization stable if and only if it has no Killing fields. Our current work extends these results and describes precisely the geometry of the space of solutions of Einstein's equations in a neighborhood of a solution that has a single Killing vector field. In particular, in a neighborhood of such a spacetime, the solutions *cannot* be parameterized in a smooth way by elements of a linear space, such as the space of four functions of three variables. (See Barrow and Tipler [8] for an application of this result.)

We shall begin by describing a theorem from bifurcation theory and give a simple example that is a prototype for what is happening in relativity.

Theorem 1   Let $\phi: \mathbb{R}^n \to \mathbb{R}^k$ be a smooth mapping satisfying $\phi(0) = 0$ and $D\phi(0) = 0$ (i.e., the matrix $\partial\phi^i/\partial x^j$ of partial derivatives of ϕ vanishes at the origin). Let

$$Q(v) = D^2\phi(0) \cdot (v, v) = \frac{\partial^2\phi^i}{\partial x^j \partial x^l}(0)v^i v^l$$

and suppose that whenever $Q(v) = 0$ and $v \neq 0$, the matrix

$$a^i_j = \frac{\partial^2\phi^i}{\partial x^j \partial x^l}(0)v^l$$

has rank k, i.e., the map $w \mapsto D^2\phi(0)(v, w)$ is onto. Then the set of solutions of

$$\phi(x) = 0 \tag{1}$$

for x near 0 is homeomorphic to the cone of solutions of

$$Q(v) = 0. \tag{2}$$

Notice that Eq. (2) is a set of k simultaneous homogeneous quadratic equations in the n variables $v^1, \ldots, v^n$. This theorem states that to determine the nature of the solutions of (1), one may restrict one's attention to the second-order terms in the Taylor expansion of ϕ about 0; i.e., one may disregard terms higher than the second order. Thus, roughly, we may say that singularities of this type are "conical" since they are determined by the second-order terms of their Taylor series.

The above result is proved by a method called "blowing up a singularity," wherein one considers the scaled equation

$$\tilde{\phi}(x, r) = r^{-2} \cdot \phi(rx) = 0 \qquad \text{for a real variable } r \text{ and a unit vector } x$$

in $\mathbb{R}^n$. This change of variables expands the origin to the unit sphere. In these new variables, the implicit function theorem may be applied to deduce the theorem above (see Buchner *et al.* [9] and Marsden [10]).

Some generalizations of Theorem 1 are important for general relativity. For example, if ϕ is invariant under the action of a group G, one may be able to construct a manifold N of solutions of (1) by exploiting this symmetry property. The above theorem may then be applicable if ϕ is restricted to variables transverse to N.

The following well-known theorem illustrates a variant of Theorem 1 for $k = 1$ but in the presence of symmetry groups (see Bott [11]). The theorem is a consequence of the Morse lemma.

Theorem 2   Let $\phi : \mathbb{R}^n \to \mathbb{R}$ be smooth and satisfy

$$\phi(0) = 0, \qquad D\phi(0) = 0.$$

Suppose ϕ has a nondegenerate critical manifold $N \subset \mathbb{R}^n$ through 0; i.e., N is a submanifold on which $D\phi = 0$ and $D^2\phi(0)$ restricted to a space transverse to $T_0 N$ (its tangent space at the origin) is nonsingular. Then the conclusions of the previous theorem apply; in particular, the solutions of $\phi(x) = 0$ near N have the structure of a product of a cone with N.

As a simple example, consider the structure of the set of solutions (x, y, z) to the equation

$$\phi(x, y, z) = x^2 + y^2 - 2(x^2 + y^2)^{1/2} - z^2 + 1 = 0$$

near the solution $(0, 1, 0)$. Since ϕ is rotationally invariant about the z axis, the circle of radius 1 in the xy plane, labeled N in Fig. 1, is also a curve of solutions. If ϕ is restricted to variables transverse to N at $(0, 1, 0)$, i.e., if ϕ is restricted to the yz plane, then the above theorem applies, showing that the solution set in these transverse variables is a cone in the yz plane with vertex $(0, 1, 0)$. Due to the rotational invariance of ϕ, the full solution set in $\mathbb{R}^3$ is then given by the circle N from each point of which a cone C of solutions branches, as shown in Fig. 1.

For relativity, the previous theorems need to be generalized to allow ϕ to be defined on an infinite dimensional space. To study spacetimes near one with one Killing field, it suffices to consider a real-valued function on an infinite-dimensional space, as in Theorem 2. For spacetimes with k-Killing

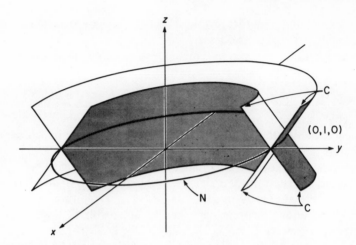

Fig. 1   A cone C of solutions of $\phi(x, y, z) = 0$ branching off a manifold N of solutions.

fields, an $\mathbb{R}^k$-valued function on an infinite-dimensional space is involved. The precise statement is rather technical and we refer to Buchner *et al.* [9] for the statement. The general methods, however, are reflected in the above theorems.

Now we apply these ideas to the empty space Einstein equations

$$\mathrm{Ein}(^{(4)}g) = 0, \tag{3}$$

where $\mathrm{Ein}(^{(4)}g) = \mathrm{Ric}(^{(4)}g) - \frac{1}{2}{}^{(4)}gR(^{(4)}g)$. We shall assume that our space-times $(V_4, {}^{(4)}g)$ are globally hyperbolic with compact Cauchy hypersurfaces Σ. Let (g, π) be the Cauchy data induced on Σ by the spacetime $(V_4, {}^{(4)}g)$, so that g is a Riemannian metric on Σ with canonically conjugate momentum density $\pi = \pi' \, d\mu_g$ (π' is the "tensor part" of π), and $d\mu_g$ is the volume element associated with g.

Let $(V_4, {}^{(4)}g)$ be a globally hyperbolic Einstein flat spacetime, $\mathrm{Ein}(^{(4)}g) = 0$. Then $(V_4, {}^{(4)}g)$ is a *maximal spacetime* if it is the maximal development of some triple (Σ, g, π), where Σ is a Cauchy surface with Cauchy data (g, π) (see Hawking and Ellis [12, p. 249], for the definition of a maximal development). Note that if a spacetime is maximal with respect to a Cauchy surface Σ, it is then maximal with respect to any other Cauchy surface. For M compact, and $V_4 = \mathbb{R} \times M$, we let

$$\mathscr{E}_{\max} = \mathscr{E}_{\max}(V_4)$$
$$= \{^{(4)}g \,|\, \mathrm{Ein}(^{(4)}g) = 0, \text{ and } (V_4, {}^{(4)}g) \text{ is a maximal spacetime}\}.$$

By the initial value theorems of general relativity, solutions to the Einstein equations are determined up to coordinate transformations by solutions of the constraint equations

$$\Phi(g, \pi) = (\mathscr{H}(g, \pi), \mathscr{J}(g, \pi)) = 0 \qquad (4)$$

on Σ, where

$$\mathscr{H}(g, \pi) = (\pi' \cdot \pi' - \tfrac{1}{2}(\operatorname{tr} \pi')^2 - R(g))d\mu_g,$$

$$\mathscr{J}(g, \pi) = 2\delta_g \pi.$$

Here $R(g)$ is the scalar curvature of g, and $\delta_g \pi$ is the divergence of π (see Arnowitt *et al.* [13]).

The Eqs. (4) are a powerful "representation" of the Einstein equations (3). This is so because the Fréchet derivative $D\Phi(g, \pi)$ has an elliptic adjoint $D\Phi(g, \pi)^*$, whereas the Einstein equations are hyperbolic in nature. Thus, through the constraint equations (4), methods, techniques, and theorems concerning elliptic operators may be brought to bear on the study of the Einstein equations. (See Berger and Ebin [14] for a discussion of some of these elliptic methods.)

Symmetries of Eq. (3) are reflected in (4) as follows. If $^{(4)}X$ is a vector field on V_4 with normal and tangential components $(^{(4)}X_\perp, {}^{(4)}X_\parallel)$ to Σ, then $^{(4)}X$ is a Killing vector field for $^{(4)}g$ only if $(^{(4)}X_\perp, {}^{(4)}X_\parallel)$ lies in ker $D\phi(g, \pi)^*$.

Using these ideas Fischer and Marsden [1–3], Moncrief [4–6], and Arms and Marsden [7] have previously proved the following.

Theorem 3   The space of solutions $\mathscr{E}_{\max}$ is a smooth infinite-dimensional manifold in a neighborhood of a solution $^{(4)}g_0 \in \mathscr{E}_{\max}$ with tangent space the space of solutions of the linearized equations if and only if $^{(4)}g_0$ has no Killing vector fields.

According to this theorem, near a spacetime $^{(4)}g_0 \in \mathscr{E}_{\max}$ with symmetries, $\mathscr{E}_{\max}$ is not a manifold with its natural tangent space. Our current study shows that such singular regions in the space of solutions contain conical structures analogous to those in Fig. 1. In other words, $\mathscr{E}_{\max}$ branches, or bifurcates, in a neighborhood of spacetimes with Killing vector fields.

This branching of solutions is closely related to the phenomenon of linearization instability, but provides much more information about the structure of the space of solutions to Einstein's equations. We say that Eqs. (3) are *linearization stable* at $^{(4)}g_0 \in \mathscr{E}_{\max}$ if for any solution $^{(4)}h$ of the linearized Einstein equations

$$D \operatorname{Ein}(^{(4)}g_0) \cdot {}^{(4)}h = 0 \qquad (5)$$

and compact set $D \subset V_4$, there is a curve $^{(4)}g(\rho)$ of exact solutions of (3) on D such that on D, $^{(4)}g(0) = {}^{(4)}g_0$ and $d^{(4)}g(\rho)/d\rho|_{\rho=0} = {}^{(4)}h$; i.e., $^{(4)}h$ is *integrable*.

If there exists some nonintegrable $^{(4)}h$ that is a solution of (5), then we say that the Einstein equations (2) are *linearization unstable* at $^{(4)}g_0$.

If $^{(4)}g_0$ has a Killing vector field $^{(4)}X$, we can find a necessary quadratic condition in order that a solution $^{(4)}h$ of (5) be integrable. The development of this second-order condition is based on the discovery of a second-order conserved quantity due to Taub [15,16]. We first establish two lemmas, which we shall refer to as the Taub lemmas.

Lemma 4 *If* $\text{Ein}(^{(4)}g) = 0$, *and* $^{(4)}h$ *is any symmetric two tensor, then*

$$\delta_{(4)g}(D \, \text{Ein}(^{(4)}g) \cdot {}^{(4)}h) = 0,$$

where $\delta_{(4)g}$ *denotes the covariant divergence with respect to* $^{(4)}g$.

Proof The contracted Bianchi identities assert that

$$\delta_{(4)g} \, \text{Ein}(^{(4)}g) = 0.$$

Differentiation then gives the identity

$$(D\delta(^{(4)}g) \cdot {}^{(4)}h) \cdot \text{Ein}(^{(4)}g) + \delta_{(4)g}(D \, \text{Ein}(^{(4)}g) \cdot {}^{(4)}h) = 0,$$

where $\delta(^{(4)}g)$ indicates the functional dependence of $\delta_{(4)g}$ on $^{(4)}g$, and $D\delta(^{(4)}g) \cdot {}^{(4)}h$ is the derivative of this function. The lemma follows since $\text{Ein}(^{(4)}g) = 0$. ∎

Lemma 5 *Suppose* $\text{Ein}(^{(4)}g) = 0$ *and* $D \, \text{Ein}(^{(4)}g) \cdot {}^{(4)}h = 0$. *Then*

$$\delta_{(4)g}(D^2 \, \text{Ein}(^{(7)}g) \cdot ({}^{(4)}h, {}^{(4)}h)) = 0.$$

Proof Differentiate the contracted Bianchi identity $\delta_{(4)g} \, \text{Ein}(^{(4)}g) = 0$ twice with respect to $^{(4)}g$ and argue as in the preceding lemma. ∎

Putting together Taub's lemmas gives us *Taub's Theorem*.

Theorem 6 *Suppose* $\text{Ein}(^{(4)}g) = 0$, $D \, \text{Ein}(^{(4)}g) \cdot {}^{(4)}h = 0$, *and* $^{(4)}X$ *is a Killing vector field for* $^{(4)}g$. *Then*

$$^{(4)}T = {}^{(4)}X \cdot (D^2 \, \text{Ein}(^{(4)}g) \cdot ({}^{(4)}h, {}^{(4)}h))$$

has zero divergence (i.e., $^{(4)}T$ *is a conserved quantity; here "·" denotes contraction).*

Proof

$$\delta_{(4)g}({}^{(4)}X \cdot (D^2 \, \text{Ein}(^{(4)}g) \cdot ({}^{(4)}h, {}^{(4)}h)))$$
$$= \tfrac{1}{2}(L_{(4)X} \, {}^{(4)}g) \cdot (D^2 \, \text{Ein}(^{(4)}g) \cdot ({}^{(4)}h, {}^{(4)}h))$$
$$+ {}^{(4)}X \cdot \delta_{(4)g}(D^2 \, \text{Ein}(^{(4)}g) \cdot ({}^{(4)}h, {}^{(4)}h))$$

so the result follows from the hypothesis $L_{(4)X}$ $^{(4)}g = 0$ and Lemma 5.   ■

In particular, for two compact spacelike hypersurfaces Σ_1 and Σ_2,

$$\int_{\Sigma_1} {}^{(4)}T \cdot {}^{(4)}Z_{\Sigma_1} \, d\mu_{\Sigma_1} = \int_{\Sigma_2} {}^{(4)}T \cdot {}^{(4)}Z_{\Sigma_2} \, d\mu_{\Sigma_2},$$

where $^{(4)}Z_{\Sigma_i}$ is the unit forward pointing timelike normal to Σ_i, and $d\mu_{\Sigma_i}$ is the Riemannian volume element induced on Σ_i by $^{(4)}g$. Thus Taub's conserved quantity

$$B(\Sigma, {}^{(4)}h, {}^{(4)}X) = \int_{\Sigma} {}^{(4)}T \cdot {}^{(4)}Z_{\Sigma} \, d\mu_{\Sigma}$$

is independent of the hypersurface on which it is evaluated.
    We shall also need the following related result:

Lemma 7   Suppose $\mathrm{Ein}(^{(4)}g) = 0$, $^{(4)}X$ *is a Killing field of* $^{(4)}g$, *and* $^{(4)}h$ *is a symmetric two-tensor field, and* Σ *is a compact spacelike hypersurface. Then*

$$I(\Sigma, {}^{(4)}h) = \int_{\Sigma} \langle {}^{(4)}X \cdot (D \, \mathrm{Ein}(^{(4)}g) \cdot {}^{(4)}h), {}^{(4)}Z \rangle \, d\mu_{\Sigma} = 0.$$

Proof   By Lemma 4, $D \, \mathrm{Ein}(^{(4)}g) \cdot {}^{(4)}h$ has zero divergence, and so the vector field $^{(4)}X \cdot (D \, \mathrm{Ein}(^{(4)}g) \cdot {}^{(4)}h)$ also has zero divergence. Thus for any two compact spacelike hypersurfaces Σ_1 and Σ_2,

$$I(\Sigma_1, {}^{(4)}h) = I(\Sigma_2, {}^{(4)}h).$$

Choose Σ_1 and Σ_2 disjoint and replace $^{(4)}h$ by a symmetric two-tensor $^{(4)}k$ that equals $^{(4)}h$ on a tubular neighborhood of Σ_1 and vanishes on a tubular neighborhood of Σ_2. Then

$$I(\Sigma_1, {}^{(4)}h) = I(\Sigma_1, {}^{(4)}k) = I(\Sigma_2, {}^{(4)}k) = 0.   ■$$

These ideas are connected to linearization stability by the following result (see Fischer and Marsden [2,3] and Moncrief [6]):

Theorem 8   Suppose $\mathrm{Ein}(^{(4)}g_0) = 0$, $^{(4)}X$ *is a Killing vector field of* $^{(4)}g_0$, *and* $^{(4)}h$ *is an integrable solution to the linearized equations. Then the conserved quantity of Taub vanishes identically when integrated over any compact spacelike hypersurface* Σ,

$$B(\Sigma, {}^{(4)}X) = \int_{\Sigma} \langle {}^{(4)}X \cdot (D^2 \, \mathrm{Ein}(^{(4)}g_0) \cdot ({}^{(4)}h, {}^{(4)}h)), {}^{(4)}Z_{\Sigma} \rangle \, d\mu_{\Sigma} = 0. \quad (6)$$

Proof   Let $^{(4)}g(\lambda)$ be a curve of exact solutions through $^{(4)}g_0$ and tangent to $^{(4)}h$ at $^{(4)}g_0$. Differentiating $\text{Ein}(^{(4)}g(\lambda)) = 0$ twice with respect to λ and evaluating at $\lambda = 0$ gives the identity

$$D^2 \, \text{Ein}(^{(4)}g_0) \cdot (^{(4)}h, \, ^{(4)}h) + D \, \text{Ein}(^{(4)}g_0) \cdot \, ^{(4)}k = 0,$$

where

$$^{(4)}k = \frac{d^2}{d\lambda^2} \, g(\lambda)|_{\lambda=0}$$

is the "acceleration" of $^{(4)}g(\lambda)$ at $\lambda = 0$. Contracting with $^{(4)}X$ and integrating over Σ gives

$$\int_\Sigma \langle \, ^{(4)}X \cdot (D^2 \, \text{Ein}(^{(4)}g_0) \cdot (^{(4)}h, \, ^{(4)}h)), \, ^{(4)}Z_\Sigma \rangle d\mu_\Sigma$$

$$+ \int_\Sigma \langle \, ^{(4)}X \cdot (D \, \text{Ein}(^{(4)}g_0) \cdot \, ^{(4)}k), \, ^{(4)}Z_\Sigma \rangle d\mu_\Sigma = 0$$

The last integral vanishes by Lemma 7. ∎

This theorem is important when a spacetime has Killing vector fields, since it provides a necessary second-order condition in order that a first-order deformation $^{(4)}h$ be integrable. Thus unless the conserved quantity of Taub vanishes, $B(\Sigma, \, ^{(4)}h, \, ^{(4)}X) = 0$, the first-order solution $^{(4)}h$ cannot be tangent to any curve of exact solutions. Theorem 9 below states that the vanishing of Taub's conserved quantity is also sufficient for integrability of a first-order deformation $^{(4)}h$. Thus for spacetimes that have Killing vector fields, and hence are not linearization stable, the numerical value of Taub's conserved quantity plays the central role in testing whether or not perturbations $^{(4)}h$ are actually tangent to a curve of exact solutions.

In terms of the constraint equations, the second-order condition becomes

$$\int_\Sigma \langle (^{(4)}X_\perp, \, ^{(4)}X_\parallel), D^2\Phi(g, \pi) \cdot ((h, \omega), (h, \omega)) \rangle d\mu_g = 0 \qquad (7)$$

where (h, ω) is the perturbation of (g, π) induced by $^{(4)}h$. These quadratic conditions (6) or (7) are analogous to Equations (2) in the bifurcation analysis sketched earlier. Indeed, the main result of our current work is that this bifurcation analysis applies to the Einstein equations (3) via the constraint equations (4). In terms of linearization stability, our main result may be stated as a converse to Theorem 8.

Theorem 9   Let $(V_4, \, ^{(4)}g)$ be a maximal spacetime such that $\text{Ein}(^{(4)}g) = 0$. Assume that $(V_4, \, ^{(4)}g)$ has a compact Cauchy hypersurface with $\text{tr } \pi' = \text{constant}$, and that $^{(4)}g$ has a one-dimensional space of Killing vector fields.

Let $^{(4)}h$ satisfy the linearized equations

$$D \operatorname{Ein}(^{(4)}g) \cdot {}^{(4)}h = 0.$$

Then $^{(4)}h$ is integrable if and only if $^{(4)}h$ satisfies the second-order condition

$$\int_{\Sigma} \langle {}^{(4)}X \cdot (D^2 \operatorname{Ein}(^{(4)}g) \cdot (^{(4)}h, {}^{(4)}h)), {}^{(4)}Z_{\Sigma} \rangle d^3\Sigma = 0$$

on any Cauchy hypersurface Σ.

Remarks

(1)   A similar result holds when the solution $^{(4)}g$ has more than one Killing vector field. The precise details of this more general case are now under investigation.

(2)   The above results should hold for a variety of systems coupled to gravity, such as the Einstein–Maxwell system, or the Einstein–Yang–Mills system. (See Arms [18,19] for linearization stability results regarding these systems.)

(3)   It is believed that the requirement of a hypersurface with constant mean curvature is not very restrictive (see Choquet-Bruhat *et al.* [20]).

Theorem 9 states that in the presence of a Killing vector field the second-order condition is not only necessary but is also sufficient for integrability of first-order deformations. It is the sufficiency of this second-order quadratic condition that tells us that the singular regions in the solution space $\mathscr{E}_{\max}$ are at worst singularities of a conical type, and that the second-order condition above defines the tangent direction to the conical singularity. As has been pointed out by Moncrief [21], it is important to take these singularities into account in the quantum theory of gravity.

In order to understand the geometry behind the bifurcations that are taking place, it is necessary to isolate the directions of degeneracy analogous to the manifold N in the finite dimensional example shown in Fig. 1. For general relativity these degeneracy directions occur not only because of the coordinate covariance of the Einstein equations, but also because there will be nonisometric spacetimes nearby with one Killing vector field.

Since we are dealing with spacetimes with only one Killing vector field and which satisfy the tr $\pi' = $ constant condition, there are two cases to analyze in the proof of Theorem 9; the case when $^{(4)}X$ is spacelike, and the case when $^{(4)}X$ is timelike.

We shall comment on the case when $^{(4)}X$ is timelike later. First, suppose that $^{(4)}X$ is spacelike. In this case, the conical singularity that occurs in $\mathscr{E}_{\max}$ in a neighborhood of a solution $^{(4)}g$ is closely related to the purely geometric conical singularity that occurs in the divergence equation

$$\mathscr{J}(g, \pi) = 2\delta_g \pi = 0 \tag{8}$$

near a solution (g, π) that has precisely one Killing vector field X; i.e., a vector field X such that

$$L_X g = 0 \qquad \text{and} \qquad L_X \pi = 0.$$

An important ingredient in the analysis of the singularity that occurs in (8) about such a (g, π) is the Ebin–Palais slice theorem (see Ebin [22]) extended to a contangent bundle action. We now describe this analysis.

Let $\mathcal{M} = \text{Riem}(M)$ denote the space of Riemannian metrics on a compact manifold M, and let $\mathcal{D} = \text{Diff}(M)$ denote the group of diffeomorphisms of M. Then $\mathcal{D}$ acts on $\mathcal{M}$ by pull-back, and $\mathcal{D}$ lifts naturally to a symplectic action on the L_2-cotangent bundle $T^*\mathcal{M}$;

$$\mathcal{D} \times T^*\mathcal{M} \to T^*\mathcal{M},$$

$$(f, (g, \pi)) \mapsto (f^*g, f^*\pi),$$

where $f^*\pi = (f^{-1})_*\pi$ is the pull-back of contravariant tensor densities.

For $(g, \pi) \in T^*\mathcal{M}$, let

$$\psi_{(g, \pi)} : \mathcal{D} \to T^*\mathcal{M} ; f \mapsto (f^*g, f^*\pi)$$

denote the orbit map through (g, π), and let $\mathcal{O}_{(g, \pi)} = \text{Image } \psi_{(g, \pi)} = \{(g', \pi') \in T^*\mathcal{M} \,|\, g' = f^*g, \ \pi' = f^*\pi \text{ for some } f \in \mathcal{D}\}$ denote the orbit through (g, π).

Let $\mathcal{J} : T^*\mathcal{M} \to \Lambda_d^1 ; (g, \pi) \mapsto \mathcal{J}(g, \pi) = 2\delta_g \pi$ denote the divergence map, and let

$$J = \begin{pmatrix} 0 & I \\ -I & 0 \end{pmatrix} : S_d^2 \times S_2 \to S_2 \times S_d^2 ;$$

$$\begin{pmatrix} \omega \\ h \end{pmatrix} \mapsto J \begin{pmatrix} \omega \\ h \end{pmatrix} = \begin{pmatrix} h \\ -\omega \end{pmatrix}$$

be the symplectic matrix on $T^*\mathcal{M}$. Here Λ_d^1 is the space of one-form densities on M; S_2 is the space of symmetric two-covariant tensor fields on M; S_d^2 is the space of symmetric two-contravariant tensor densities on M; and $\mathcal{X}$ is the space of vector fields on M.

The derivatives of $\psi_{(g, \pi)}$, $\mathcal{J}$, and their natural L_2 adjoints are related as follows:

Lemma 10

$$T_{id}\psi_{(g, \pi)} = J \circ (D\mathcal{J}(g, \pi))^* : \mathcal{X} \to S_2 \times S_d^2 ; X \mapsto (L_X g, L_X \pi),$$

$$(T_{id}\psi_{(g, \pi)})^* = D\mathcal{J}(g, \pi) \circ J^* = -D\mathcal{J}(g, \pi) \circ J : S_d^2 \to \Lambda_d^1 ;$$

$$(\omega, h) \mapsto D\mathcal{J}(g, \pi) \cdot \begin{pmatrix} -h \\ \omega \end{pmatrix}.$$

Moreover $S_2 \times S_d^2$ splits L_2 orthogonally as

$$S_2 \times S_d^2 = \text{range } T_{id}\psi_{(g,\pi)} \oplus (\ker(T_{id}\psi_{(g,\pi)})^*)^*$$
$$= \text{range } J \circ (D\mathcal{J}(g,\pi))^* \oplus (\ker(D\mathcal{J}(g,\pi) \circ J))^*$$

where ( ) denotes the L_2 adjoint map*

$$(\ )^*: S_d^2 \times S_2 \to S_2 \times S_d^2 ; (\omega, h) \mapsto ((\omega^\flat)', h^\sharp d\mu_g).$$

Proof   (Here $\omega^\flat$ is the covariant form of ω and $h^\sharp$ the contravariant form of
h.) $D\mathcal{J}(g,\pi)^* \cdot X = (-L_X\pi, L_Xg)$. Hence $J \circ D\mathcal{J}(g,\pi)^* \cdot X = (L_Xg, L_X\pi) = (T_{id}\psi_{(g,\pi)}) \cdot X$, and thus $(T_{id}\psi_{(g,\pi)})^* = D\mathcal{J}(g,\pi) \circ J^* = -D\mathcal{J}(g,\pi) \circ J$ since
$J^* = -J$. The splitting follows from injectivity of the symbol of $D\mathcal{J}(g,\pi)^*$
(see Fischer and Marsden [17]).   ∎

We shall also need the following result:

Lemma 11   *The orbit $\mathcal{O}_{(g,\pi)}$ through (g,π) is a closed submanifold of $T^*\mathcal{M}$*
with tangent space at (g,π) given by

$$T_{(g,\pi)}\mathcal{O}_{(g,\pi)} = \text{range } J \circ D\mathcal{J}(g,\pi)^*$$
$$= \{(L_Xg, L_X\pi) | X \text{ is a vector field on } M\}.$$

Proof   Let $\Psi: \mathcal{D} \mapsto T^*\mathcal{M}; \eta \mapsto (\eta^*g, \eta^*\pi)$. We have

$$T_\eta\Psi(X) = (\eta^*L_Xg, \eta^*L_X\pi).$$

Since $X \mapsto L_Xg$ is elliptic, $T_\eta\Psi$ has closed range and finite-dimensional
kernel. By the arguments of Ebin and Marsden ([23], Appendix B), $\ker T_\eta\Psi$
is a subbundle of $T\mathcal{D}$. It follows from the implicit function theorem that the
range of Ψ is an immersed submanifold. From Ebin ([22], Proposition
6.13), it follows that Ψ is an open map onto its range and that the range is
closed. The lemma then follows.   ∎

The proof of our results depends on a carefully constructed slice for the
action of $\mathcal{D}$ on $T^*\mathcal{M}$. For the action of $\mathcal{D}$ on $\mathcal{M}$, the Ebin–Palais slice
theorem (see Ebin [22]) asserts the existence of a slice. To avoid unnecessary
technicalities, our slice will be an affine one, L_2 orthogonal to the orbit of $\mathcal{D}$.
Let

$$I_g = \{f \in \mathcal{D}(M) | f^*g = g\},$$

$$I_\pi = \{f \in \mathcal{D}(M) | f^*\pi = \pi\},$$

and

$$I_{(g,\pi)} = I_g \cap I_\pi.$$

Since the isometry group I_g is a compact finite-dimensional Lie group, and
since I_π is a closed subgroup of $\mathcal{D}(M)$, $I_{(g,\pi)} = I_g \cap I_\pi$ is also a compact

finite-dimensional Lie group. $I_{(g,\pi)}$ is the isotropy group at (g, π) for the action of $\mathscr{D}$ on $T^*\mathscr{M}$.

Theorem 12   The action

$$\mathscr{D} \times T^*\mathscr{M} \to T^*\mathscr{M};$$

$$(f, (g, \pi)) \mapsto (f^*g, f^*\pi),$$

has a slice $S_{(g,\pi)} \subset T^*\mathscr{M}$ at each $(g, \pi) \in T^*\mathscr{M}$; i.e., $S_{(g,\pi)}$ is a submanifold of $T^*\mathscr{M}$ containing (g, π) such that

(i)   *if $f \in I_{(g,\pi)}$, then $f^*(S_{(g,\pi)}) = S_{(g,\pi)}$;*
(ii) *if $f \in \mathscr{D}$ and $f^*(S_{(g,\pi)}) \cap S_{(g,\pi)} \neq \phi$, then $f \in I_{(g,\pi)}$;*

and

(iii)   *there is a local cross section $\chi: \mathscr{D}/I_{(g,\pi)} \to \mathscr{D}$.*

defined in a neighborhood U of the identity coset such that the map $(\phi, (g', \pi')) \mapsto (\chi(\phi))^*(g', \pi')$ is a homeomorphism of $U \times S_{(g,\pi)}$ onto a neighborhood of (g, π) in $T^*\mathscr{M}$. In particular the slice $S_{(g,\pi)}$ sweeps out a neighborhood of (g, π) under the group action.

The tangent space to the slice at (g, π) is given by

$$T_{(g,\pi)}S_{(g,\pi)} = (\ker(T_{id}\psi_{(g,\pi)})^*)^* = (\ker(D\mathscr{J}(g, \pi) \circ J))^*.$$

Sketch of the Slice Construction

Fix (g, π) and let G denote the L_2-metric on $T^*\mathscr{M}$ given by

$$G_{(g',\pi')}((h, \omega), (h, \omega)) = \int_M (\langle h, h \rangle_g + \langle \omega', \omega' \rangle_g)d\mu_g$$

for $(h, \omega) \in T_{(g',\pi')}T^*\mathscr{M} = S_2 \times S_d^2$, where $\langle, \rangle$ indicates contraction using g (note that the inner product is independent of (g', π')).

The splitting of $T_{(g,\pi)}T^*\mathscr{M}$ defined by Lemma 10 is, by construction, an orthogonal splitting with respect to G. We may exponentiate the subspace $(\ker(D(g, \pi) \circ J))^*$ to obtain an "affine" submanifold $A_{(g,\pi)} = \{(g, \pi)\} + [\ker(D\mathscr{J}(g, \pi) \circ J)]^*$ of $T^*\mathscr{M}$, which intersects $\mathscr{O}_{(g,\pi)}$ orthogonally at (g, π). The action of $I_{(g,\pi)}$ on $T^*\mathscr{M}$ leaves $A_{(g,\pi)}$ invariant. This follows from the fact that $I_{(g,\pi)}$ is an isometry group of G. We may now intersect $A_{(g,\pi)}$ with a sufficiently small ball (in a Sobolev norm invariant under $I_{(g,\pi)}$) to obtain a slice $S(g, \pi)$ for the action of $\mathscr{D}$ on $T^*\mathscr{M}$.

The proof that such "affine slices" are indeed slices now follows from the methods of Ebin and Palais.

An important consequence of the existence of a slice for the action of $\mathscr{D}$ on $T^*\mathscr{M}$ is the local decreasing property of the groups $I_{(g,\pi)}$:

Corollary 13 Let $(g, \pi) \in T^*\mathcal{M}$, and let $S_{(g, \pi)}$ be a slice at (g, π). Then if $(g', \pi') \in S_{(g, \pi)}$,

$$I_{(g', \pi')} \subseteq I_{(g, \pi)}.$$

Proof The inclusion follows from property (ii) of a slice. ∎

Let $B_{(g, \pi)} = \{(g', \pi') \in S_{(g, \pi)} | I_{(g', \pi')} = I_{(g, \pi)}\}$. Thus $B_{(g, \pi)}$ is the set of elements of the slice with the same symmetry group $I_{(g, \pi)}$. In the terminology of superspace (Fischer [24]), $B_{(g, \pi)}$ is a local stratum of (g, π)s with the same symmetry type.

Let $I^0_{(g, \pi)}$ denote the connected component of the identity of the Lie group $I_{(g, \pi)}$ and let $\mathcal{I}_{(g, \pi)}$ denote its Lie algebra. Since $\mathcal{O}_{(g, \pi)}$ is closed, the discrete part $I_{(g, \pi)}/I^0_{(g, \pi)}$ of $I_{(g, \pi)}$ never accumulates on $\mathcal{O}_{(g, \pi)}$. Since $B_{(g, \pi)} \subset S_{(g, \pi)}$, $B_{(g, \pi)}$ can also be described using $I^0_{(g, \pi)}$ rather than the full isotropy group $I_{(g, \pi)}$; i.e., if $S_{(g, \pi)}$ is sufficiently small, then

$$B_{(g, \pi)} = \{(g', \pi') \in S_{(g, \pi)} | I^0_{(g', \pi')} = I_{(g, \pi)}\}$$
$$= \{(g', \pi') \in S_{(g, \pi)} | \mathcal{I}_{(g', \pi')} = \mathcal{I}_{(g, \pi)}\}.$$

Proposition 14 $B_{(g, \pi)}$ is a submanifold of $S_{(g, \pi)}$ with tangent space at (g, π) given by

$T_{(g, \pi)} B_{(g, \pi)}$
$$= \{(h, \omega) \in (\ker(D\mathcal{I}(g, \pi) \circ J))^* | f^*h = h, f^*\omega = \omega \quad \text{for all } f \in I^0_{(g, \pi)}\}$$
$$= \{(h, \omega) \in (\ker(D\mathcal{I}(g, \pi) \circ J))^* | L_X h = 0, L_X \omega = 0 \quad \text{for all } X \in \mathcal{I}_{(g, \pi)}\}.$$

We shall use the submanifold $B_{(g, \pi)}$ to study the set

$$\mathcal{C}_\delta = \{(g, \pi) \in T^*\mathcal{M} | \delta_g \pi = 0\}.$$

As is shown in Fischer and Marsden [3], if $(g, \pi) \in \mathcal{C}_\delta$ satisfies $\mathcal{I}_{(g, \pi)} = 0$, then $\mathcal{C}_\delta$ is a submanifold in a neighborhood of (g, π). In such a neighborhood the slice $S_{(g, \pi)}$ is a cross section for the action of $\mathcal{D}$ on $T^*\mathcal{M}$, and $S_{(g, \pi)} \cap \mathcal{C}_\delta$ is also a submanifold of $\mathcal{C}_\delta$.

To study the structure of $\mathcal{C}_\delta$ and $S_{(g, \pi)} \cap \mathcal{C}_\delta$ when $\mathcal{I}_{(g, \pi)} \neq 0$, we shall utilize the bifurcation analysis sketched out earlier. The result in the case $\dim \mathcal{I}_{(g, \pi)} = 1$ is the following:

Theorem 15 Let $(g, \pi) \in \mathcal{C}_\delta$, let $S_{(g, \pi)}$ be a slice at (g, π) and let $B_{(g, \pi)} = \{(g', \pi') \in S_{(g, \pi)} | I_{(g', \pi')} = I_{(g, \pi)}\}$. Assume that $\dim \mathcal{I}_{(g, \pi)} = 1$; i.e., (g, π) has a single Killing vector field X, a vector field on M such that $L_X g = 0$ and $L_X \pi = 0$.
    Then

(a) $N = \mathcal{C}_\delta \cap B_{(g, \pi)}$ is a submanifold of the set $\mathcal{C}_\delta \cap S_{(g, \pi)}$;
(b) $\mathcal{C}_\delta \cap S_{(g, \pi)}$ is a product $C \times N$, where C is a cone; and
(c) $\mathcal{C}_\delta$ is locally homeomorphic to $C \times N \times (\mathcal{D}/I_{(g, \pi)})$.

(See Fig. 2.)

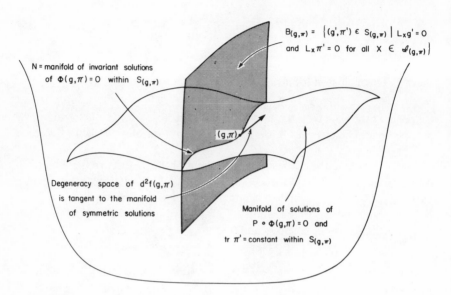

$$B_{(g,\pi)} = \left\{ (g',\pi') \in S_{(g,\pi)} \mid L_X g' = 0 \right.$$
$$\left. \text{and } L_X \pi' = 0 \text{ for all } X \in \mathscr{S}_{(g,\pi)} \right\}$$

N = manifold of invariant solutions
of $\Phi(g,\pi) = 0$ within $S_{(g,\pi)}$

(g,π)

Degeneracy space of $d^2f(g,\pi)$
is tangent to the manifold
of symmetric solutions

Manifold of solutions of
$P \circ \Phi(g,\pi) = 0$ and
$\text{tr } \pi' = \text{constant}$ within $S_{(g,\pi)}$

Fig. 2 The structure of the constraint manifold $\mathscr{C}$ in a neighborhood of a (g, π) with a Killing vector field X. The space of the picture is the slice $S_{(g,\pi)}$.

Sketch of Proof

As in the proof of Lemma 10, Λ_d^1 splits L_2 orthogonally as

$$\text{range } D\mathscr{J}(g, \pi) \oplus \ker(D\mathscr{J}(g, \pi))^*.$$

Let P denote the orthogonal projection onto the first factor. Since $D\mathscr{J}(g, \pi)^* \cdot X = (-L_X\pi, L_X g)$, the second factor is one dimensional, and is spanned by $\{X\}$. Assume X has L_2 length one and identify $\ker D\mathscr{J}(g, \pi)^*$ with the real line $\mathbb{R}$. Thus the projection $I-P$ is given by

$$(I - P) \circ \mathscr{J}(g, \pi) = \int_M X \cdot \mathscr{J}(g, \pi).$$

Let

$$\mathscr{C}_P = \{(g, \pi) \mid P \circ \mathscr{J}(g, \pi) = 0\}.$$

The map $\mathscr{J}$ is clearly transverse to $\ker D(g, \pi)^*$ and its kernel splits, so that in a neighborhood of (g, π), it follows from the implicit function theorem that $\mathscr{C}_P$ is a smooth manifold with tangent space

$$T_{(g, \pi)}\mathscr{C}_P = \ker D\mathscr{J}(g, \pi),$$

the space of solutions to the linearized constraint equation.

Let $f: \mathscr{C}_P \to \mathbb{R}$ be defined by

$$f(g, \pi) = (I - P) \circ \mathscr{J}(g, \pi)$$

$$= \int_M X \cdot \mathscr{J}(g, \pi) = \int_M \pi \cdot L_X g = 2 \int_M X \cdot \delta_g \pi.$$

Clearly the constraint set $\mathscr{C}_\delta = \mathscr{J}^{-1}(0)$ is given by

$$\mathscr{C}_\delta = f^{-1}(0).$$

Thus our problem is to analyze the zero set of f near (g, π). In bifurcation theory, the above construction of P, $\mathscr{C}_P$, and f is called the Liapunov–Schmidt procedure (see Marsden [10]).

For $(h, \omega) \in S_2 \times S_d^2$,

$$df(g, \pi) \cdot (h, \omega) = \int_M \omega \cdot L_X g + \int_M \pi \cdot L_X h = \int_M \omega \cdot L_X g - \int_M L_X \pi \cdot h.$$

Thus (g, π) is a critical point of f if and only if $L_X g = 0$ and $L_X \pi = 0$; i.e., if and only if $X \in \ker D\mathscr{J}(g, \pi)^*$. At a critical point, the Hessian of f may be computed in the ambient space $T^* \mathscr{M}$ and then restricted to $\mathscr{C}_P$. Thus

$$d^2 f(g, \pi) \cdot ((h_1, \omega_1), (h_2, \omega_2)) = \int_M \omega_1 \cdot L_X h_2 + \int_M \omega_2 \cdot L_X h_1.$$

Note that $d^2 f$ is independent of (g, π).

Since $B_{(g,\pi)} = \{(g', \pi') \in S_{(g,\pi)} | \mathscr{J}_{(g',\pi')} = \mathscr{J}_{(g,\pi)}\}$, $B_{(g,\pi)}$ is a critical submanifold of f. Moreover, it is easy to see that the degeneracy of the bilinear form $d^2 f(g, \pi)$ is tangent to this submanifold of critical points intersected with the constraint set,

$$N = \mathscr{H}_\delta \cap B_{(g,\pi)}.$$

Thus on a submanifold transverse to N, $d^2 f$ is weakly nondegenerate.

Within this setting a generalization of Theorem 2 to infinite dimensions applies and gives the structure of $\mathscr{C}_\delta \cap S_{(g,\pi)}$ as the product of a (manifold) $\times$ (cone), where the cone directions at (g, π) are given by the solutions (h, ω) of the second-order condition

$$d^2 f(g, \pi) \cdot ((h, \omega), (h, \omega)) = 2 \int \omega \cdot L_X h = 0.$$

For the map $\mathscr{J}$ restricted to $S_{(g,\pi)}$, the set $N = \mathscr{C}_\delta \cap B_{(g,\pi)}$ plays the role of N in Theorem 2. ∎

Remark

When $k = \dim \mathscr{I}_{(g, \pi)} > 1$, the set $\mathscr{C}_\delta \cap S_{(g, \pi)}$ will have the structure of "cones on cones."

The result of Theorem 15 can be generalized to give the structure of the constraint equations

$$\Phi(g, \pi) = (\mathscr{H}(g, \pi), \mathscr{J}(g, \pi)) = 0$$

in a neighborhood of $(g, \pi) \in \mathscr{C} = \mathscr{C}_\mathscr{H} \cap \mathscr{C}_\delta$, which satisfies the hypotheses

(i) $\dim \mathscr{I}_{(g, \pi)} = 1$; and
(ii) $\operatorname{tr} \pi' = \mathrm{const.}$

The conclusion is identical to (a)–(c) of Theorem 15 with $\mathscr{C}$ replacing $\mathscr{C}_\delta$; see Fig. 2. Thus the conical singularities in $\mathscr{C}_\delta$ are carried over to $\mathscr{C}$. Using this result on the structure of $\mathscr{C}$, we can now sketch a proof of Theorem 9 (see [25] for details).

Sketch of Proof of Theorem 9 We have already remarked on the necessity of the second-order condition for integrability of first-order deformations $^{(4)}h$. To prove sufficiency, we first consider the case that $^{(4)}X$ is spacelike. From the fact that Σ has constant mean curvature, one can show that $^{(4)}X$ is parallel to Σ, so that the perpendicular–parallel decomposition of $^{(4)}X$ along Σ is $(^{(4)}X_\perp, {}^{(4)}X_\|) = (0, X)$, where X is now a vector field on the submanifold Σ (identified with M).

Let (g, π) be the Cauchy data induced on Σ by $^{(4)}g$. In Moncrief [4], it is proven that

$$\ker(D\Phi(g, \pi)^*) = \{({}^{(4)}X_\perp, {}^{(4)}X_\|) | {}^{(4)}X \in \mathscr{I}_{(4)_g}\},$$

where $\mathscr{I}_{(4)_g}$ is the space of Killing vector fields for $^{(4)}g$. In the case at hand, $^{(4)}g$ has only a single spacelike Killing vector field, so $\ker(D\Phi(g, \pi))^*$ is the one-dimensional space spanned by $(0, X)$. Moreover,

$$D\Phi(g, \pi)^* \cdot (0, X) = D\mathscr{J}(g, \pi)^* \cdot X = (-L_X\pi, L_X g),$$

so that if $(0, X) \in \ker D\Phi(g, \pi)^*$, then $L_X g = 0$ and $L_X \pi = 0$.

Let $^{(4)}g_0 \in \mathscr{E}_{\max}$ be a maximal development of (Σ, g, π), so that $(^{(4)}g, V_4) \subseteq (^{(4)}g_0, V_4)$. Let $I_{(4)_{g_0}}$ denote the isometry group of $^{(4)}g_0$ and let

$$B_{(4)_{g_0}} = \{{}^{(4)}g' \in \mathscr{E}_{\max} | I_{(4)_g} = F^{-1} \circ I_{(4)_{g_0}} \circ F \quad \text{for some } F \in \mathrm{Diff}(V_4)\},$$

i.e., $B_{(4)_{g_0}}$ is the set of solutions to the field equations with the same *symmetry type* as $^{(4)}g_0$. Thus $B_{(4)_{g_0}}$ includes all solutions $^{(4)}g'$ related to $^{(4)}g_0$ by a transformation F; $^{(4)}g' = F^{*(4)}g_0$, but it also includes other nonisometric solutions as well.

Let P be the projection to range $D\Phi(g, \pi)$, $\mathscr{C}_P = \{(\bar{g}, \bar{\pi}) | P\Phi(\bar{g}, \bar{\pi}) = 0\}$ and define the function

$$f : \mathscr{C}_P \cap S_{(g, \pi)} \to \mathbb{R}$$

by

$$f(h, \omega) = \int_{\Sigma} (^{(4)}X_{\perp}, {}^{(4)}X_{\parallel}) \cdot \Phi(g + h, \pi + \omega)$$

$$= \int_{\Sigma} {}^{(4)}X_{\perp} \cdot (g + h, \pi + \omega) + \int_{\Sigma} X \cdot \mathscr{J}(g + h, \pi + \omega)$$

$$= \int_{\Sigma} X \cdot \mathscr{J}(g + h, \pi + \omega),$$

since $X_{\perp} = 0$. Applying the methods used in Theorem 15, one can show that the degeneracy subspace of the second derivative of this function at (g, π) is described by the Cauchy data on Σ corresponding to elements of the manifold $B_{(4)g_0}$, and that conical structures appear in the transverse direction. Thus, in a neighborhood of $^{(4)}g_0$, the solution space $\mathscr{E}_{\max}$ looks like a set of cones bifurcating off of the manifold $B_{(4)g_0}$ as is depicted in Fig. 1. Thus, locally, in a neighborhood of $^{(4)}g_0$, $\mathscr{E}_{\max}$ has the product structure (manifold) $\times$ (cone) where the manifold is $B_{(4)g_0}$ and the cone C consists of the zeros $^{(4)}h$ of the second-order condition (6), representing solutions with less symmetry. Thus the directions of the cone field are delineated precisely by this second-order condition.

A similar argument can be given for the case of one timelike Killing vector field. In this case the solutions with the same symmetry are all stationary and hence all flat. As in the spacelike case, a cone of solutions of lower symmetry branches from each flat spacetime. ∎

The conical structure when $^{(4)}g_0$ has just one Killing vector field can be visualized as follows. Consider an open book. The spine of the book corresponds to the manifold $B_{(4)g_0}$ and the sides of the book correspond to the conical structures bifurcating off $B_{(4)g_0}$. Going along the spine of the book corresponds to moving in $\mathscr{E}_{\max}$ in a direction that preserves the symmetry type of $^{(4)}g_0$, whereas moving up the sides of the book in the conical direction corresponds to breaking the symmetry and moving to generic empty space solutions of the Einstein equations that do not have any symmetry.

The results of our analysis show that the space of solutions of Einstein's equations has an unexpected complexity near a spacetime with symmetries. The complexity points out the cautions required when doing perturbation analysis and the dangers of function counting arguments. It is hoped that these results shed light on the nature of solutions of Einstein's equations near spacetimes with Killing vector fields, their stability as regards the breaking of these symmetries, and corresponding difficulties that occur with the quantum theory. Since our own Universe apparently is highly symmetrical in the large, the structure of the space of solutions near spacetimes with Killing vector fields is a question of some importance.

References

[1]   Fischer, A., and Marsden, J., *Bull. Am. Math. Soc.* **79**, 997–1003 (1973).
[2]   Fischer, A., and Marsden, J. *Proc. Symp. Pure Math.*, *Am. Math. Soc.* **27**, Part 2, 219–263 (1975).
[3]   Fischer, A., and Marsden, J., *in* "Isolated Gravitating Systems in General Relativity" (J. Ehlers, ed.), pp. 322–395 North-Holland Publ., New York and Amsterdam, 1979.
[4]   Moncrief, V., *J. Math. Phys.* **16**, 493–498 (1975).
[5]   Moncrief, V., *J. Math. Phys.* **16**, 1556–1560 (1975).
[6]   Moncrief, V., *J. Math. Phys.* **17**, 1893–1902 (1976).
[7]   Arms, J., and Marsden, J., *Indiana Univ. Math. J.* **28**, 119–125 (1979).
[8]   Barrow, J., and Tipler, F., *Phys. Reports* **56**, 371–420 (1979).
[9]   Buchner, M., Marsden, J., and Schecter, S., to be published (1980).
[10] Marsden, J., *Bull. Am. Math. Soc.* **84**, 1125–1148 (1978).
[11] Bott, R., *Ann. Math.* **60**, 248–261 (1954).
[12] Hawking, S. W., and Ellis, G. F. R., "The Large Scale Structure of Spacetime." Cambridge Univ. Press, London and New York, 1973.
[13] Arnowitt, R., Deser, S., and Misner, C. W., *in* "Gravitation: An Introduction to Current Research" (L. Witten, ed.), pp. 227–265. Wiley, New York, 1962.
[14] Berger, M., and Ebin, D., *J. Differ. Geom.* **3**, 379–392 (1969).
[15] Taub, A. H., *J. Math. Phys.* **2**, 787–793 (1961).
[16] Taub, A. H., *in* "Relativistic Fluid Dynamics" (C. Cattaneo, ed.), pp. 205–300. Edizioni Cremonese, Rome, 1971.
[17] Fischer, A., and Marsden, J., *Duke Math. J.* **42**, 519–547 (1975).
[18] Arms, J., *J. Math. Phys.* **18**, 830–833 (1977).
[19] Arms, J., *J. Math. Phys.* **20**, 443–453 (1979).
[20] Choquet-Bruhat, Y., Fischer, A., and Marsden J., *in* "Isolated Gravitating Systems in General Relativity" (J. Ehlers, ed.), pp. 396–456, North-Holland Publ., New York and Amsterdam, 1972.
[21] Moncrief, V., *Phys. Rev.* **18**, 983–989 (1978).
[22] Ebin, D., *Proc. Symp. Pure Math.*, *Am. Math. Soc.* **15**, 11–40 (1970).
[23] Ebin, D., and Marsden, J., *Ann. Math.* **92**, 102–163 (1970).
[24] Fischer, A., *in* "Relativity" (M. Carmeli, S. Fickler, and L. Witten, eds.), pp. 303–357. Plenum, New York, 1970.
[25] Fischer, A., Marsden, J., and Moncrief, V., "The structure of the space of solutions of Einstein's equation. I: One Killing field," to appear (1980).

8

Gauge Invariant Perturbation Theory in Spatially Homogeneous Cosmology

Robert T. Jantzen†

Department of Physics and Astronomy
University of North Carolina
Chapel Hill, North Carolina

Abstract

Lie group harmonic analysis is applied to the solution of tensor equations on LRS class A spatially homogeneous spacetimes. The techniques developed are used to discuss the Hamiltonian dynamics of the linearized vacuum Einstein equations and Moncrief's orthogonal decomposition of the linearized phase space adapted to the linearized gauge transformations. This leads to what Moncrief has called "gauge invariant perturbation theory" for the class of spacetimes under consideration.

I. Introduction

In relativity, it is especially important to have examples that illustrate the ideas and techniques of the general theory. Spatially homogeneous cosmology, introduced by Taub several decades ago [1], has served a very important role in providing a wealth of examples with which to investigate various aspects of the theoretical perspective that has evolved in relativity over the intervening years. Misner, Ryan and Ellis, among many others, have used this class of spacetimes to study such things as Hamiltonian dynamics, singularities, and global structure. Taub participated in these

† Permanent address: 8 Houston Street, Florida, New York 10921.

97

applications at crucial points, most notably being the study of the Taub–Nut spacetimes [2] and the important clarification of the role of variational principles in spatially homogeneous cosmology [3].

Recently, it has been recognized that spatially homogeneous cosmology provides extremely nice examples [4] of many of the ideas connected with the gauge aspects of relativity as developed by Fischer and Marsden [5], Moncrief [6], Smarr and York [7,8], and others. The various quotient spaces ("true minisuperspace," "true minisuperphasespace," and their conformal analogs) and their relation to the Hamiltonian dynamics, the "true degrees of freedom" of the gravitational field (at the "group level") and the geometrical slicing techniques [7] of Smarr and York (at the "Lie algebra level") is particularly clear in this finite-dimensional context that lends very convincing support for the latter techniques. Because one has such a strong handle on the spatial aspects of the three-plus-one picture for these spacetimes, one can actually reduce the Hamiltonian dynamics by the three-dimensional diffeomorphism group, thus eliminating the momentum constraints and three kinematical degrees of freedom (in the nondegenerate case). Using the Hamiltonian constraint to eliminate an additional kinematical degree of freedom associated with the conformal factor, one obtains (in the nondegenerate case) a reduced Hamiltonian system for two gravitational degrees of freedom describing the conformal three-geometry of the spatially homogeneous slices. In this process, the minimal distortion shift vector field of Smarr and York plays a key role. Furthermore, additional symmetries of the spatially homogeneous spacetimes lead to nontrivial isotropy groups on submanifolds of the configuration and phase spaces and singular points in the quotient spaces, leading finally to linearization instabilities in a finite-dimensional setting exhibiting all of the features discussed by Fischer and Marsden [5] and Moncrief [6] in the general case.

During the past decade, Moncrief has developed the Hamiltonian form of the linearized Einstein equations based on earlier work by Taub [9] and an approach that adapts that variational approach to the "gauge group" of relativity, i.e., gauge invariant perturbation theory [10,10a]. This is related to his orthogonal decomposition of the linearized gravitational phase space and reflects the linearized version of the true degrees of freedom and quotient space questions [6]. When the background spacetime has symmetries, conserved quantities quadratic in the perturbations exist (originally investigated by Taub [9]), which generate the action of the symmetry group on the linearized system and which play an important role in linearization stability theory when the spacelike slices are compact [6]. However, there exist no nontrivial applications of this formalism or nontrivial examples of linearization instabilities. (Moncrief's treatment of spherical symmetry uses an oblique decomposition obtained by hand [9].) Naturally, spatially homogeneous cosmology has again offered itself for illustrative purposes. In this paper we shall consider the perturbations of the vacuum LRS spatially homogeneous spacetimes from a Hamiltonian viewpoint and examine the Moncrief decomposition of the canonical variables and its relation to gauge invariant perturbation theory [11]. This requires the development of some preliminary mathematical machinery associated with local rotational symmetry.

II. Spatially Homogeneous Spacetimes

Regardless of the source, the spatially homogeneous spacetimes for which eigentensor expansion techniques are useful in perturbation theory can be divided up into the following four intersecting categories:

(i)   general abelian case (Bianchi type I);
(ii) general compact case (Bianchi type IX);
(iii) locally rotationally symmetric (LRS) class A case; and
(iv) isotropic case (positive, zero, and negative curvature Friedmann models).

Only in the first three cases and their intersection with the last is Lie group harmonic analysis useful in analyzing vacuum or nonvacuum perturbations.

In the type I case (which contains an LRS model and the flat Friedmann models), ordinary Fourier analysis applies to the components of tensor fields in cartesian coordinates on the flat spacelike slices; the linearized Einstein equations reduce to ordinary differential equations (ODEs) on each space of fixed Fourier coefficients of the components of all the fields involved in the perturbation problem. (Perko *et al.* have used this approach to treat the dust case [12].) In the type IX case (which contains an LRS model and in the nonvacuum case, the positive curvature Friedmann models), generalized Fourier analysis for compact Lie groups applies, and the linearized equations reduce to ODEs on each finite-dimensional irreducible subspace of the "right tensor dragging representation" associated with the collection of fields that are involved. (Hu has examined the vacuum models using this technique [13].) In the bi-invariant type IX case (positive curvature Friedmann models), Lie group techniques may be used to develop a natural set of tensor harmonics that reduce the linearized equations to ODEs on each space of fixed "tensor Fourier coefficients" of all the fields of the problem [14]. These harmonics and their analytic continuation to the negative curvature case were implicitly used by Lifshitz in his classic treatment of the perturbations of the nonzero curvature perfect fluid Friedmann models [15] and have been recently explicitly evaluated in spherical coordinates by Gerlach and Sengupta [16].

However, in the remaining cases, due to the fact that the irreducible unitary representations of a noncompact group are infinite dimensional, Lie group harmonic analysis is of no help unless one's attention is restricted to models of higher symmetry, the additional symmetry having the effect of confining the coupling of the ODEs obtained by expansion techniques to finite-dimensional spaces whose generic dimension equals the number of independent components of all the fields involved in the perturbation problem. The only such models that are not isotropic are the LRS class A spacetimes that have a four-parameter isometry group. (We exclude the Kantowski–Sachs models and their type III analogs [17].) The vacuum models of this category consist of the type IX Taub–Nut spacetimes and their analytic continuation to type VIII and their Lie group contraction to type II, as well as the LRS Kasner models (the type I and type VII$_0$ LRS models coincide) [18,19]. All but the type VIII solutions are contained in Taub's original paper on the subject [1]. The linearized Einstein equations for the vacuum type IX and perfect fluid models of this category were studied by Bonanos without the benefit of variational principle techniques [20].

The manifold of a spatially homogeneous spacetime is of the form $M = R \times G$ where R is the real line with natural coordinate t and G is a connected three-dimensional Lie group [4]. All of the structure on each component manifold can be identified with analogous structure naturally induced on the product manifold. For example, the coordinate vector field and one-form on R and tensor fields on G or diffeomorphisms of G into itself all induce corresponding fields or diffeomorphisms on M. In particular, the Lie algebras $\mathfrak{g}$ and $\tilde{\mathfrak{g}}$ of, respectively, left and right invariant vector fields on G play important roles when considered as Lie subalgebras of $\mathscr{X}(M)$. Let $e = \{e_a\}$ and $\tilde{e} = \{e_a\}$ be corresponding bases of $\mathfrak{g}$ and $\tilde{\mathfrak{g}}$ (i.e., their values at the identity agree) with dual bases of, respectively, left and right invariant one-forms $\{\omega^a\}$ and $\{\tilde{\omega}^a\}$ and let $C^a{}_{bc} = \omega^a([e_a, e_b]) = -\tilde{\omega}^a([\tilde{e}_a, \tilde{e}_b])$ be the components of the structure constant tensor of $\mathfrak{g}$ in the basis e. The Lie brackets of $\mathfrak{g}$ with $\tilde{\mathfrak{g}}$ vanish. We assume that e is a canonical basis of $\mathfrak{g}$ in the sense of reference 4. Also introduce the Lie derivative operators $L_a = i\mathscr{L}_{e_a}$ and $L_a = -i\mathscr{L}_{e_a}$, both of which satisfy commutation relations of the form

$$[L_a, L_b] = iC^c{}_{ab} L_c. \tag{II.1}$$

We shall refer to these, respectively, as right and left angular momentum. The Lie algebra $\tilde{\mathfrak{g}}$ generates the left translations of G into itself, or when considered as a Lie subalgebra of $\mathscr{X}(M)$, the natural left action of G on M.

The Lorentzian spacetime metric ${}^4g$ on M is required to be invariant under the natural left action of G on M and to be such that the copies G_t of G in M (spatially homogeneous hypersurfaces or slices) are spacelike for values of t in some open interval of R; we shall confine our attention to such an interval when it is not all of R. Since the three-plus-one approach is so natural to spatially homogeneous spacetimes considering their natural slicing by spatially homogeneous hypersurfaces, we shall use that approach exclusively [21]. Choosing zero shift vector field, we may write

$$^4\!\mathscr{g} = -N(t)^2 \, dt \otimes dt + g_{ab}(t)\omega^a \otimes \omega^b. \tag{II.2}$$

$\mathscr{g} = g_{ab}\omega^a \otimes \omega^b$ is a time dependent left invariant metric on G. Indeed any spatially homogeneous tensor field on M may be thought of as a collection of time-dependent left invariant fields on G in a natural way. (A spatially homogeneous tensor field on M is one that is invariant under the left action of G on M.)

For the LRS spacetimes, the matrix $\mathbf{g}$ of metric components is of the form $\mathbf{g} = \text{diag}(A, A, B)$. The matrix $\pi = \text{diag}\,(\pi_A, \pi_A, \pi_B)$ of mixed components of the gravitational canonical momentum Π and the matrix $R = \text{diag}(R_A, R_A, R_B)$ of mixed components of the Ricci tensor R of $\mathscr{g}$ are also of this form, as is the matrix of components of any second-rank tensor that is left invariant and LRS like $\mathscr{g}$ itself. For a given choice of the lapse function $N(t)$, solutions $(g(t), \pi(t))$ of the Hamiltonian equations of motion determine the spacetime metric. These equations are easily integrated with the Misner choice of lapse that assumes that $2l = Ng^{1/2}A^{-1}$ is a positive constant. We use the abbreviations $g = \det \mathbf{g} = A^2 B$ and $\delta = (A/B)^{1/2}$.

The nonzero canonical structure constant tensor components for the LRS class A types are $C^1{}_{23} = C^2{}_{31} = n$, $C^3{}_{12} = \mathcal{N}$ and their antisymmetric counterparts. For the "nondegenerate" types VII_0, VIII, and IX, $n = 1$ and $\mathcal{N}$ assumes the respective values 0, -1, and 1, while $n = 0$ for the "degenerate" types I and II and $\mathcal{N}$ assumes the respective values 0 and 1. A quadratic Casimir operator for the right or left angular momentum algebras can be introduced for types IX, VIII, and VII_0. The right Casimir operator is defined by

$$L^2 = \mathcal{N}(L_1^2 + L_2^2) + L_3^2 \tag{II.3}$$

in the first two cases and in the latter case by

$$L^2 = L_1^2 + L_2^2.$$

The left Casimir operators $\tilde{L}^2$ are obtained by replacing L_a by $\tilde{L}_a$; when acting on scalar fields $L^2 = \tilde{L}^2$. For type II, $L_3 = -\tilde{L}_3$ is itself a Casimir operator, while everything commutes in the abelian type I case.

III. ξ-spin and Spherical Bases

The additional symmetry for the LRS spacetimes is associated with a one-parameter subgroup of automorphisms of G whose action on G is generated by a vector field $\xi \in \mathfrak{aut}(G)$, which induces a rotation of the canonical basis e about e_3 in a counterclockwise direction:

$$-iM_3 e_a = \mathscr{L}_\xi e_a = \varepsilon_{3ab} e_b$$

$$\xi_\theta e_a = e^{-i\theta M_3} e_a = e_b \mathscr{R}^b{}_a, \qquad \mathscr{R} = \exp \theta(e^1{}_2 - e^2{}_1). \qquad \text{(III.1)}$$

$\{e^a{}_b\}$ is the natural basis of $\mathfrak{gl}(3, R)$ consisting of matrices whose only non-zero entry is a 1 in the bth row and ath column, while $\mathscr{R}$ is the matrix of the above-mentioned rotation by an angle θ. The Killing Lie algebra of either $^4\mathscr{g}$ or $\mathscr{g}$ has $\{\tilde{e}_a, \xi\}$ as a basis. Any left invariant LRS tensor field T on G (or spatially homogeneous LRS tensor field on M) is annihilated by the corresponding Lie derivatives:

$$\tilde{L}_a T = M_3 T = 0, \qquad \text{(III.2)}$$

while any left invariant LRS tensor-valued linear differential operator on G (or corresponding object on M) must commute with these Lie derivative operators. For types VII_0, VIII, and IX, $\xi = e_3 - \tilde{e}_3$ generates an inner automorphism subgroup while $e_3 - \tilde{e}_3$ vanishes for types I and II, and ξ generates an outer automorphism subgroup. [The three Killing vector fields $\{e_3, \tilde{e}_3, \xi\}$ are not linearly independent but satisfy the relation $n(e_3 - \tilde{e}_3) = \xi$.] The manifold of this subgroup is S^1 and it is naturally isomorphic to $SO(2, R)$.

It is the existence of this one-parameter group of automorphisms acting on $(M, {}^4\mathscr{g})$ as an isometry group that permits an elegant application of Lie group harmonic analysis in the solution of tensor equations on the space-time that are invariant under the full four-dimensional isometry group. Using the canonical frame e, one may introduce a decomposition of M_3 into orbital and spin parts leading to a notion called ξ-spin very much like the third component of ordinary spin in nonrelativistic quantum mechanics and like the third component of the $SO(3, R)$ spin introduced for $SU(2)$ in Jantzen [14]. [For the type IX case, M_3 coincides with J_3 of that reference except that only the $SO(2, R)$ subgroup associated with e_3 is relevant here.]

$$M_3 = M_3^{\text{orb}} + M_3^{\text{spin}},$$

$$M_3^{\text{orb}}(T^{a\cdots}{}_{b\ldots} e_a \otimes \cdots \otimes \omega^b \cdots) = M_3(T^{a\cdots}{}_{b\ldots}) e_a \otimes \cdots \otimes \omega^b \cdots,$$

$$M_3^{\text{spin}}(T^{a\cdots}{}_{b\ldots} e_a \otimes \cdots \otimes \omega^b \cdots) = T^{a\cdots}{}_{b\ldots} M_3(e_a \otimes \cdots \otimes \omega^b \cdots). \qquad \text{(III.3)}$$

By expanding all tensor fields of G of a given type in a left invariant eigenbasis of ξ-spin (i.e., eigentensors of M_3) called a "spherical basis," we can apply to the "spherical components" (expansion coefficients in a spherical basis) Lie group harmonic analysis [22] on the space $L^2(G)$ of complex-valued functions that are square-integrable with respect to the bi-invariant Lesbesque measure associated with the bi-invariant three-form $\omega^{123} = \omega^1 \wedge \omega^2 \wedge \omega^3$ on G:

$$\langle f_1, f_2 \rangle = \int_G \omega^{123} \bar{f}_1 f_2, \quad \langle f, f \rangle < \infty, \quad f \in L^2(G) \qquad \text{(III.4)}$$

It is the fact that eigentensors of M_3 with different eigenvalues decouple in equations invariant under the full isometry group that enables one to reduce such equations to ODEs on the spaces of fixed eigenvalues of M_3 and two other commuting operators that make a complete commuting set, the dimensions of which are generically equal to the total number of independent components of all the fields involved. The situation is therefore comparable to the general abelian case as far as the harmonic analysis is concerned.

Since ξ generates rotations of e about e_3, one may introduce spherical bases that are eigentensors of M_3 by treating e as if it were a cartesian frame on R^3 and using familiar formulas for the corresponding spherical bases. For vector fields, one-forms and second-rank contravariant, covariant and mixed tensor fields, introduce, respectively, the spherical bases $\{e_A\}$, $\{\omega^A\}$, $\{e_{sm}\}$, $\{\omega^{sm}\}$, and $\{\hat{\omega}^{sm}\}$. where A and B assume the values $\pm 1, 0$ and m the values, $\pm 2, \pm 1, 0$ which label the eigenvalues of M_3, while $s = 0, 1, 2$:

$$e_0 = e_3, \qquad e_{\pm 1} = \mp 2^{-1/2}(e_1 \pm ie_2),$$

$$\omega^0 = \omega^3, \qquad \omega^{\pm 1} = \mp 2^{-1/2}(\omega^1 \pm i\omega^2),$$

$$e_{sm} = C_{11}(sm, AB)e_A \otimes e_B,$$

$$\omega^{sm} = C_{11}(sm, AB)\omega^A \otimes \omega^B,$$

$$\hat{\omega}^{sm} = C_{11}(sm, AB)e_A \otimes \omega^B. \qquad \text{(III.5)}$$

We use the caret notation to indicate the mixed form of any second-rank tensor. Explicit expressions for the Clebsch–Gordan coefficients appearing here may be found in Jantzen [14]. These complex bases all satisfy reality conditions of the form

$$\overline{e_A} = (-1)^A e_{-A}, \qquad \overline{e_{sm}} = (-1)^m e_{s-m}, \qquad \text{(III.6)}$$

and are dual to each other in the sense that

$$\overline{\omega^A}(e_B) = \omega^A(\overline{e_B}) = \delta^A_B,$$

$$C_{\overline{e_{sm}}}(\omega^{s'm'}) = C_{e_{sm}}(\overline{\omega^{s'm'}}) = \delta_{ss'}\delta_{mm'}, \qquad \text{(III.7)}$$

where C represents the natural contraction of tensors with the opposite valence. The elements of the covariant and contravariant second-rank spherical bases are symmetric for $s = 0, 2$ and antisymmetric for $s = 1$.

The metric $\mathscr{g}$ (and indeed any LRS left invariant second-rank covariant tensor) may be written in the form

$$\mathscr{g} = a\mathscr{g}_0 + b\omega^{20}, \qquad M_3\mathscr{g} = 0, \qquad (III.8)$$

where $\mathscr{g}_0 = \delta_{ab}\omega^a \otimes \omega^b = -\sqrt{3}\omega^{00}$ and a and b are certain linear combinations of A and B. Since the metric has zero ξ-spin, raising and lowering indices can only link eigentensors of fixed ξ-spin. For second-rank tensors, this results in a linear transformation on the s index on the subspace of fixed m values relative to the trivial index raising and lowering associated with the metric $\mathscr{g}_0$. It is convenient to have second-rank spherical bases that are related by index raising and lowering relative to $\mathscr{g}$ instead of $\mathscr{g}_0$. This is accomplished by introducing the second-rank spherical bases generated by treating the orthonormal frame $\{E_a\} = \{A^{-1/2}e_1, A^{-1/2}e_2, B^{-1/2}e_3\}$ as a cartesian frame. Let $\{W^a\}$ be the dual frame and $\{E_A\}$ and $\{W^A\}$ the corresponding spherical bases, defined as in Eq. (III.5). The desired second-rank spherical bases are then,

$$E_{sm} = C_{11}(sm, AB)E_A \otimes E_B,$$

$$W^{sm} = C_{11}(sm, AB)W^A \otimes W^B, \qquad (III.9)$$

$$\hat{W}^{sm} = C_{11}(sm, AB)E_A \otimes W^B.$$

The elements of each basis with $s \neq 1$ span the space of tensors that are either naturally symmetric or symmetric with respect to $\mathscr{g}$.

Suppose X, σ, and h are a vector field, one-form and symmetric covariant second-rank tensor field on G. We define their spherical components to be their expansion coefficients in the appropriate spherical basis:

$$X = X^A e_A, \qquad X^A = \overline{\omega^A}(X), \qquad \overline{X^A} = (-1)^A X_{-A},$$

$$\sigma = \sigma_A \omega^A, \qquad \sigma_A = \sigma(\overline{e_A}), \qquad \overline{\sigma_A} = (-1)^A \sigma_{-A}, \qquad (III.10)$$

$$h = h_{sm} W^{sm}, \qquad h_{sm} = C_{\overline{E_{sm}}}(h), \qquad \overline{h_{sm}} = (-1)^m h_{s-m}.$$

Note that the $s = 1$ components of h vanish and that the mixed form $\hat{h}$ or its contravariant form (relative to $\mathscr{g}$) have the same spherical components since E_{sm}, W^{sm}, and $\hat{W}^{sm}$ are related by index raising and lowering with respect to $\mathscr{g}$.

The space of mixed second-rank tensors at a point is isomorphic to the algebra of 3×3 matrices under the operation of contraction of adjacent indices (indicated by an $\llcorner$ or $\lrcorner$ pointing from the covariant index to the contravariant one). Any LRS left invariant second-rank mixed tensor field T is of the form $T = T_0 id + T_2 \hat{W}^{20}$, where $id = \delta^a{}_b e_a \otimes \omega^b = -\sqrt{3}\hat{W}^{00}$ is the identity tensor. "Multiplication" of arbitrary mixed second-rank tensors by such a tensor therefore requires knowledge of the result of left and right multiplication of $\{\hat{W}^{sm}\}$ by $\hat{W}^{20}$ (linear transformations in the index s for fixed m since $\hat{W}^{20}$ has zero ζ-spin). This multiplication table is easily evaluated. Two trace properties of the basis $\{\hat{W}^{sm}\}$ are useful:

$$\text{Tr } \hat{W}^{sm} = -\sqrt{3}\delta_{s0}\delta_{m0},$$
$$\text{Tr } \overline{\hat{W}^{sm}} \llcorner W^{s'm'} = (-1)^s \delta_{ss'}\delta_{mm'}. \tag{III.11}$$

These trace and multiplication properties enable one to evaluate any expression quadratic in two second-rank mixed tensor (or tensor density) fields, each of which may undergo linear transformations induced by LRS left-invariant second-rank mixed fields like the mixed Ricci tensor $\hat{R}$ of g and the mixed form $\hat{\pi}$ of the gravitational canonical momentum. Index raising and lowering by g is avoided by using the mixed forms of all relevant second-rank fields. The example we have in mind is the evaluation of the Hamiltonian for the perturbations of the LRS class A spatially homogeneous spacetimes in terms of spherical components.

The notion of ζ-spin can be extended to LRS left invariant tensor-valued linear differential operators on G. For example, if B and F are a one-form-valued operator and a symmetric $\binom{0}{2}$ tensor-valued operator each of this type, we may introduce their spherical components:

$$B = \omega^A B_A, \qquad B_A = C_{\bar{e}_A} \circ B, \qquad [M_3, B_A] = -AB_A,$$
$$F = W^{sm} F_{sm}, \qquad F_{sm} = C_{\overline{E_{sm}}} \circ F, \qquad [M_3, F_{sm}] = -mF_{sm}. \tag{III.12}$$

The indices A and m of B_A and F_{sm} will be referred to as their ζ-spin values. The simplest examples are the corresponding identity operators whose spherical components are just the contraction operators $C_{\bar{e}_A}$ and $C_{\overline{E_{sm}}}$ themselves which assign to a one-form σ and a $\binom{0}{2}$ tensor field h their spherical components σ_A and h_{sm}, respectively. The next simplest example is the exterior derivative acting on scalars, an LRS left-invariant one-form-valued first-order linear differential operator on scalar fields:

$$d = -i\omega^a L_a = -i\omega^A L_A, \qquad L_A = -iC_{\bar{e}_A} \circ d,$$
$$L_{\pm 1} = \mp 2^{-1/2} L_\pm, \qquad L_0 = L_3. \tag{III.13}$$

$L_\pm = L_1 \pm iL_2$ are the "raising and lowering" operators. The operators L_3 and $L_{(+}L_{-)} = L_1^2 + L_2^2$ commute with M_3 and thus have zero ζ-spin while $L_\pm$ has ζ-spin ∓ 1. ζ-spins of products of operators add. The operators B_0, $L_\pm B_{\pm 1}$, F_{so}, $L_\pm F_{s\pm 1}$, $(L_\pm)^2 F_{2\pm 2}$ all have zero ζ-spin. This indicates a way to obtain zero ζ-spin expressions from those with nonzero ζ-spin. For example, if X and h are a vector field and $\binom{0}{2}$ tensor, respectively, introduce the following "zero ζ-spin variables":

$$\mathcal{X}^0 = X^0, \qquad \mathcal{X}^{\pm 1} = L_\pm X^{\pm 1},$$
$$h_{so} = h_{so}, \qquad h_{s\pm 1} = L_\pm h_{s\pm 1}, \qquad h_{2\pm 2} = L_\pm^2 h_{2\pm 2}. \tag{III.14}$$

Using the commutation relations of $L_\pm$, L_3, and $L_{(+}L_{-)}$, the "zero ξ-spin components" of any LRS left invariant linear differential operator on vector fields X or $\binom{0}{2}$ tensors h (etc.), can be rewritten in terms of powers of L_3 and $L_{(+}L_{-)}$ acting on the "zero ξ-spin variables." (However the transformation to zero ξ-spin variables for variables with initially nonzero ξ-spin $\pm|m|$ annihilates the part of $L^2(G)$ belonging to the kernel of $(L_\pm)^{|m|}$ so these subspaces must be separately considered.)

Thus if one has a complete orthonormal basis $\{Q^j\}$ of $L^2(G)$ consisting of (possibly generalized) eigenfunctions of L_3, $L_{(+}L_{-)}$, and a third commuting operator such as $\tilde{L}_3$, L_1, or M_3, LRS spatially homogeneous tensor equations on M can be reduced to ODEs on each space of fixed "Fourier coefficients" of the "zero ξ-spin" spherical component variables of all the tensor fields involved. It is this property of ξ-spin that Bonanos used [20] without realizing the underlying structure present in the problem.

Our method will be to directly "Fourier analyze" the spherical components of the tensor (or tensor density) fields involved in the perturbation problem, thus relating different "Fourier coefficients" of different spherical components in the resulting ODEs (except in the abelian case where both methods coincide). In the Hamiltonian approach we proceed in two steps. First the Hamiltonian is explicitly evaluated in terms of the spherical components of the fields of the problem. Then the generalized expansion of each spherical component is introduced into the result and the integration is carried out using the orthogonality properties of the generalized eigenfunctions $\{Q^j\}$:

$$\langle Q^j, Q^{j'} \rangle = \delta_{jj'}, \qquad \text{(III.15)}$$

Here j essentially stands for a collection of eigenvalues of a complete set of three commuting operators and $\delta_{jj'}$ for the product of Kronecker deltas for discrete eigenvalues and delta functions for continuous eigenvalues. For example, in the type IX case j stands for the discrete eigenvalues JMN associated with L^2, L_3, and $\tilde{L}_3$, respectively, as in Jantzen [14], although for simplicity we assume a different normalization for the functions Q^{JMN} here.

The canonical variables for the Hamiltonian treatment of the vacuum perturbations are a symmetric $\binom{0}{2}$ tensor h and a symmetric $\binom{2}{0}$ tensor density p that are canonically conjugate in the sense that their components satisfy the following formal Poisson bracket relations (all other brackets vanish):

$$\{h_{ab}(x_1), p^{cd}(x_2)\} = \delta^c{}_{(a}\delta^d{}_{b)}\delta_e(x_1, x_2),$$

$$\qquad \text{(III.16)}$$

$$\int_{G_2} \delta_e(x_1, x_2)f(x_2)(\omega^{123})(x_2) = f(x_1), \qquad f \in L^2(G).$$

Transforming to spherical components, one finds that the complex scalar and scalar density variables $h_{sm} = (-1)^m \overline{h_{s-m}}$ and $\overline{p_{sm}} = (-1)^m p_{s-m}$ are canonically conjugate in the following sense:

$$\{h_{sm}(x_1), \overline{p_{sm}}(x_2)\} = \delta_{ss'}\delta_{mm'}\delta_e(x_1, x_2).$$ (III.17)

Their "generalized Fourier coefficients" then satisfy similar bracket relations:

$$h_{sm} = \sum_j h^j_{sm} Q^j, \qquad \overline{p_{sm}} = \sum_j p^j_{sm} Q^j,$$

$$h^j_{sm} = \langle Q^j, h_{sm}\rangle, \qquad p^j_{sm} = \langle Q^j, p_{sm}\rangle,$$ (III.18)

$$\{h^j_{sm}, \overline{p^j_{sm}}\} = \delta_{ss'}\delta_{mm'}\delta_{jj'}.$$

The generalized sum $\sum_j$ stands for a sum over the discrete eigenvalues and an integration over the continuous ones.

Since the linear canonical (point) transformation from (h_{ab}, p^{ab}) to (h_{sm}, p_{sm}) involves A and B and is therefore time dependent, one must add to the Hamiltonian a term that may be read off the rearrangement that takes place in the action integral:

$$p^{ab}\dot{h}_{ab} - \mathcal{HAM} = \overline{p_{sm}}\dot{h}_{sm} - (\mathcal{HAM} + \Delta\mathcal{HAM})$$

$$= \overline{p_{sm}}\dot{h}_{sm} - \mathcal{HAM}^{\text{spherical}}$$

$$\Delta\mathcal{HAM} = -2N \operatorname{Tr} \hat{p} \llcorner \hat{K} \llcorner \hat{h}$$ (III.19)

$$= Ng^{-1/2}(2\operatorname{Tr}\hat{p} \llcorner \pi \llcorner \hat{h} - \operatorname{Tr} \pi \operatorname{Tr} \hat{p} \llcorner \hat{h}).$$

Poisson brackets of the spherical variables with the integrated spherical Hamiltonian density $\text{HAM}^{\text{spherical}} = \int_G \omega^{123} \mathcal{HAM}^{\text{spherical}}$ then give their equations of motion.

This naive derivation of $\Delta\mathcal{HAM}$ is valid only because the linear canonical transformation is a "point transformation." (The specific formula for $\Delta\mathcal{HAM}$ above is characteristic of the point transformation associated with the normalization of the orthogonal frame $\{e_a\}$, namely, the intermediate canonical transformation $(h_{ab}, p^{ab}) \to (\tilde{h}_{ab} = h(E_a, E_b), \tilde{p}^{ab} = p(W^a, W^b)) \to (h_{sm}, p_{sm})$, which contains all the time dependence.) For a general time-dependent linear canonical transformation:

$$(Z^A) = (q^i, p_i) \to (z^{\bar{A}}) = (q^{\bar{i}}, p_{\bar{i}}), \quad Z^A = C^A{}_{\bar{B}}z^{\bar{B}}$$ (III.20)

of the canonical coordinates of the phase space of a finite-dimensional mechanical system, one can easily verify that adding the term

$$\Delta H = \tfrac{1}{2}Z^{\bar{A}}C^C_{\bar{A}}J_{CD}\dot{C}^D{}_{\bar{B}}Z^{\bar{B}}, \qquad J_{CD} = \{Z^C, Z^D\}$$ (III.21)

to the Hamiltonian H compensates for the time dependence of the transformation. In other words the Hamiltonian equations for (Z^A) following from the Hamiltonian $H + \Delta H$ are the transforms of the original Hamiltonian equations for (Z^A) following from H. We shall need this result in a subsequent section.

IV. Linearized Hamiltonian for Vacuum LRS Spatially Homogeneous Spacetimes

Moncrief has written down the Hamiltonian form of the linearized vacuum Einstein equations, based on ideas of Taub regarding linearized variational principles [6,10,10a]. In our case the formal action integral with zero shift expressed in the frame $\{\partial/\partial t, e_a\}$ on M is

$$\text{ACTION}(h, p, N', \mathbf{N}'; g, \pi, N; t_1, t_2) = \int_{[t_1;t_2] \times G} dt \wedge \omega^{123} \mathscr{L}, \quad \text{(IV.1)}$$

$$\mathscr{L} = p^{ab}\dot{h}_{ab} - \mathscr{HAM},$$

$$\mathscr{HAM} = \tfrac{1}{2}D^2\mathscr{H} + N'\mathscr{H}' + \mathbf{N}'^a\mathscr{H}'_a,$$

$$D^2\mathscr{H} = D^2\mathscr{H}(g, \pi) \cdot ((h, p), (h, p)),$$

$$\mathscr{H}' = D\mathscr{H}(g, \pi) \cdot (h, p),$$

$$\mathscr{H}'_a = \mathscr{J}'_a = D\mathscr{J}_a(g, \pi) \cdot (h, p)$$

$$= -2p^b_{a|b} - 2\pi^{bc}(h_{ab|c} - \tfrac{1}{2}h_{bc|a}),$$

$$\mathscr{H}_a = \mathscr{J}_a = -2\pi^b_{a|b}.$$

Explicit expressions for these quantities may be found in the cited work of Moncrief. We include the expressions for $\mathscr{J}_a$ and $\mathscr{J}'_a$ to emphasize that the momentum constraint functional $\mathscr{J} = \mathscr{J}_a\omega^a$ and its linearization $\mathscr{J}'$ should be considered as one-form-valued rather than vector field-valued densities in order to be dual to the shift vector field and its linearization and avoid needless complications arising from the "unnatural choice."

For example, if X is a vector field on G the following formula holds:

$$\tfrac{1}{2}X \lrcorner D^2\mathscr{J} = \tfrac{1}{2}X^a D^2\mathscr{J}_a = -2X^a(h^c_a p^b_{c|b} + \tfrac{1}{2}p^{bc}(2h_{ab|c} - h_{bc|a}))$$

$$= p^{ab}(\mathscr{L}_X h)_{ab} - 2(X^a h_{ac} p^{cb})_{|b}. \quad \text{(IV.2)}$$

Comparison with the formula:

$$X \lrcorner \mathscr{J} = \pi^{ab}(\mathscr{L}_X \mathscr{g})_{ab} - 2(X^a g_{ac} p^{cb})_{|b} \qquad \text{(IV.3)}$$

shows that the formal linearized Poisson brackets of functionals of the form

$$E(X) = \int_G \tfrac{1}{2} X \lrcorner D^2 \mathscr{J} \omega^{123} \qquad \text{(IV.4)}$$

are identical with the formal Poisson brackets [5] of the functionals

$$\langle X, \mathscr{J} \rangle = \int_G X \lrcorner \mathscr{J} \omega^{123}$$

of the full constraints, namely:

$$\{E(X), E(Y)\} = E([X, Y]). \qquad \text{(IV.5)}$$

In fact if we restrict our attention to square-integrable fields (fields whose components in the frame e are square-integrable), then the integrated divergences vanish and we have simply:

$$E(X) = \int_G \omega^{123} p^{ab}(\mathscr{L}_X h)_{ab}. \qquad \text{(IV.6)}$$

This functional generates the natural action on the linearized L^2-phase space of the one-parameter group of diffeomorphisms of G associated with the vector field X. In particular, the quantities

$$\tilde{E}_a = E(\tilde{e}_a) = i\langle p, \tilde{L}_a h \rangle,$$
$$E_\xi = E(\xi) = -i\langle p, M_3 h \rangle,$$
$$E_3 = E(e_3) = -i\langle p, L_3 h \rangle, \qquad \text{(IV.7)}$$
$$\langle p, h \rangle = \int_G \omega^{123} p^{ab} h_{ab}$$

generate the action of the isometry group of $\mathscr{g}$ on this space. These are conserved quantities since $\{\tilde{e}_a, \xi, e_3\}$ are spacetime Killing vector fields. This means that for a given solution of the linearized Hamiltonian equations at a fixed background solution, these quantities have time-independent values. In the compact case they must be constrained to vanish because of linearization stability requirements [5,6]. Except in the type I case, because of the relation

$$nM_3 = L_3 + \tilde{L}_3, \tag{IV.8}$$

at most two linearly independent operators from the set $\{\tilde{L}_a, M_3, L_3\}$ can be simultaneously diagonalized. Whether we choose $\{L_3, \tilde{L}_3, M_3\}$, $\{L_3, \tilde{L}_1\}$ or any other subset of commuting operators, then on a simultaneous eigenspace of pairs of eigentensors (h, p) with coinciding eigenvalues, the corresponding conserved quantities are each proportional to the single quantity

$$E = -i\langle p, h \rangle. \tag{IV.9}$$

[For continuous eigenvalues E will contain a factor $\delta(0)$ since the corresponding generalized eigentensors are not themselves square-integrable but satisfy delta function normalizations as in (III.15).]

The nonderivative terms in the Hamiltonian must first be written in terms of the mixed tensor (and tensor density) fields $\hat{h}$, $\hat{p}$, $\hat{\pi}$, $\hat{R}$, which permits an easy evaluation in terms of spherical components as discussed in the previous section; the term $\Delta \mathcal{H}\mathcal{A}\mathcal{M}$ which changes the Hamiltonian density to the spherical one must also be added. The first term $\frac{1}{2}ND^2\mathcal{H}$ in the Hamiltonian density together with the additional term $\Delta \mathcal{H}\mathcal{A}\mathcal{M}$ consists of a kinetic energy term $\mathcal{T}^{\text{spherical}} = \mathcal{T} + \Delta \mathcal{H}\mathcal{A}\mathcal{M}$ containing the momenta and a potential energy term $\mathcal{U}$ quadratic in h. $\mathcal{U}$ itself consists of a nonderivative part involving π and $\hat{R}$, which may be simplified somewhat using the super-Hamiltonian constraint $\mathcal{H} = 0$, and a complicated derivative term $Ng^{1/2}$ Der(h, h). Since N and $g^{1/2}$ depend only on the time t, we may integrate by parts on G treating them as constants. Furthermore, the exact differential three-forms arising in such an integration by parts in the expression $\mathcal{H}\mathcal{A}\mathcal{M}^{\text{spherical}}{}_{(i)}{}^{123}$ may be discarded without affecting the equations of motion. We can therefore replace Der(h, h) by the following expression differing only by a divergence:

$$h^{ab}(\text{DER} \cdot h)_{ab} = \frac{1}{2}h^{ab}(h_{ca|b}{}^{|c} - \frac{1}{2}h_{ab|c}{}^{|c} - \frac{1}{2}h^c{}_{c|ab})$$
$$- \frac{1}{4}h^c_c(h^{ab}{}_{|ab} - h^b{}_{b|a}{}^{|a}).$$

The LRS left invariant symmetric $\binom{0}{2}$ tensor-valued second-order linear differential operator DER defined here is formally symmetric in the sense that $k^{ab}(\text{DER} \cdot h)_{ab} - h^{ab}(\text{DER} \cdot k)_{ab}$ is a covariant divergence. (DER is a self-adjoint operator when acting on square-integrable tensor fields.) From now on we assume that $\mathcal{H}\mathcal{A}\mathcal{M}^{\text{spherical}}$ has undergone these modifications.

To evaluate the spherical Hamiltonian and associated equations of motion requires the individual evaluation of $\mathcal{T}^{\text{spherical}}$, $\mathcal{U}$, $\mathcal{H}'$, $\mathcal{J}'$, $D\mathcal{H}^*$, and $D\mathcal{J}^*$ in terms of spherical components, the most lengthly and tedious calculation being $h^{ab}(\text{DER} \cdot h)_{ab}$, which involves the complicated covariant derivatives.

The results are as follows:

$$\mathscr{H}\mathscr{A}\mathscr{M}^{\text{spherical}} = Ng^{-1/2}\{|p_{2\pm2}|^2 + |p_{2\pm1}|^2 + |p_{20}|^2$$
$$- \tfrac{1}{2}|p_{00}|^2 + \pi_A(\overline{p_{2\pm2}}h_{2\pm2} + \overline{h_{2\pm2}}p_{2\pm2})$$
$$+ \tfrac{1}{2}(\pi_A + \pi_B)(\overline{p_{2\pm1}}h_{2\pm1} + \overline{h_{2\pm1}}p_{2\pm1}) + \tfrac{1}{3}(\pi_A + 2\pi_B)(\overline{p_{20}}h_{20} + \overline{h_{20}}p_{20})$$
$$+ \tfrac{1}{12}(2\pi_A + \pi_B)(\overline{p_{00}}h_{00} + \overline{h_{00}}p_{00})$$
$$- \tfrac{1}{6}2^{1/2}(\pi_A - \pi_B)(\overline{p_{20}}h_{00} + \overline{p_{00}}h_{20} + \overline{h_{00}}p_{20} + \overline{h_{20}}p_{00})\}$$
$$+ \tfrac{1}{2}Ng^{1/2}A^{-1}\{h_{2\pm2}(2Ag^{-1}\pi_A^2 + 2n^2\delta^2 + \tfrac{1}{2}\delta^2 L_3^2$$
$$+ (2n\delta^2 - \tfrac{1}{2}\mathscr{N})L_3)h_{2\pm2} + \overline{h_{2\pm1}}(2Ag^{-1}\pi_A\pi_B + 2n\mathscr{N} + \tfrac{1}{4}L_{(+}L_{-)}$$
$$+ \tfrac{3}{4}\mathscr{N}L_3)h_{2\pm1} + \tfrac{1}{3}\overline{h_{20}}(2Ag^{-1}(2\pi_A\pi_B + \pi_B^2) + \mathscr{N}(4n + \mathscr{N}\delta^{-2})$$
$$+ L_{(+}L_{-)} - \tfrac{1}{2}\delta^2 L_3^2)h_{20} + \tfrac{1}{6}\overline{h_{00}}(Ag^{-1}(12\pi_A^2 + 20\pi_A\pi_B + \pi_B^2)$$
$$+ 4\mathscr{N}(4n - \mathscr{N}\delta^{-2}) - 2(L_{(+}L_{-)} + \delta^2 L_3^2))h_{00}$$
$$+ \tfrac{1}{6}2^{1/2}\overline{h_{20}}(Ag^{-1}(6\pi_A^2 - 5\pi_A\pi_B - \pi_B^2) - \mathscr{N}(n - \mathscr{N}\delta^{-2})$$
$$+ \tfrac{1}{2}L_{(+}L_{-)} - \delta^2 L_3^2)h_{00} + \tfrac{1}{6}2^{1/2}\overline{h_{00}}(Ag^{-1}(6\pi_A^2 - 5\pi_A\pi_B - \pi_B^2)$$
$$- \mathscr{N}(n - \mathscr{N}\delta^{-2}) + \tfrac{1}{2}L_{(+}L_{-)} - \delta^2 L_3^2)h_{20}$$
$$+ \tfrac{3}{4}n[\overline{h_{2\pm2}}L_{\mp}h_{2\pm1} + \overline{h_{2\pm1}}L_{\pm}h_{2\pm2}]$$
$$\pm \tfrac{1}{2}\delta[\overline{h_{2\pm2}}L_{(3}L_{\pm)}h_{2\pm1} + \overline{h_{2\pm1}}L_{(3}L_{\pm)}h_{2\pm2}]$$
$$+ 6^{-1/2}[((\tfrac{1}{4}n\delta + 2\mathscr{N}\delta^{-1})\overline{h_{20}} + \tfrac{1}{4}(n\delta - \mathscr{N}\delta^{-1})\overline{h_{00}})L_{\pm}h_{2\pm1}$$
$$+ \overline{h_{2\pm1}}L_{\mp}((\tfrac{1}{4}n\delta + 2\mathscr{N}\delta^{-1})h_{20} + \tfrac{1}{4}(n\delta - \mathscr{N}\delta^{-1})\sqrt{2}h_{00})]$$
$$\pm \tfrac{1}{2}\delta6^{-1/2}[(\overline{h_{20}} + 2^{1/2}\overline{h_{00}})L_{(3}L_{\pm)}\overline{h_{2\pm1}} + h_{2\pm1}L_{(3}L_{\mp)}(h_{20} + 2^{1/2}h_{00})]$$
$$+ \tfrac{1}{4}\overline{h_{2\pm1}}L_{\mp}^2h_{2\pm1} + \tfrac{1}{4}[\overline{h_{2\pm2}}L_{\mp}^2Y_h + Y_hL_{\pm}^2h_{2\pm2}]\} + \overline{N'}\mathscr{H}' + \overline{N'^A}\mathscr{I}'_A;$$

$$\mathscr{H}' = g^{1/2}A^{-1}(\tfrac{1}{2}L_{\pm}^2h_{2\pm2} - \tfrac{1}{2}(n\delta - \mathscr{N}\delta^{-1})L_{\pm}h_{2\pm1} \mp \delta L_{(3}L_{\pm)}h_{2\pm1})$$
$$+ 6^{-1/2}g^{1/2}A^{-1}(-L_{(+}L_{-)}(h_{20} - 2\,2^{1/2}h_{00}) + 2\delta^2 L_3^2(h_{20} + 2^{1/2}h_{00})$$
$$- (2\mathscr{N}(n - \mathscr{N}\delta^{-2}) + 2Ag^{-1}\pi_B(\pi_A - \pi_B))h_{20})$$
$$+ 6^{-1/2}g^{-1/2}(-4(\pi_A - \pi_B)p_{20} + (2\pi_A + \pi_B)2^{1/2}p_{00});$$

$$\mathscr{I}'_{\pm1} = \pm i2^{1/2}(-L_{\pm}p_{2\pm2} - (\mathscr{N}\delta^{-1} - \delta(\pm L_3 + n))p_{2\pm1} + L_{\mp}X_P$$
$$- \pi_A L_{\pm}h_{2\pm2} - (\pi_A\mathscr{N}\delta^{-1} - \pi_B\delta(\pm L_3 + n))h_{2\pm1} + \tfrac{1}{2}\pi_B L_{\mp}Y_h);$$

$$\mathscr{I}'_0 = i(\mp \delta^{-1}L_{\pm}(p_{2\pm1} + \pi_A h_{2\pm1}) + L_3(2Y_p + \pi_B Y_h + 2\pi_A X_h));$$

$$((h, p), D\mathscr{H}^*N') = g^{1/2}A^{-1}(\tfrac{1}{2}\overline{h_{2\pm2}}L_{\mp}^2 - \overline{h_{2\pm1}}(\tfrac{1}{2}(n\delta - \mathscr{N}\delta^{-1})L_{\mp}$$
$$\pm \delta L_{(3}L_{\mp)}) + 6^{-1/2}\overline{h_{20}}(-L_{(+}L_{-)} + 2\delta^2 L_3^2 + 2\mathscr{N}(\mathscr{N}\delta^{-2} - n)$$
$$- 2Ag^{-1}\pi_B(\pi_A - \pi_B) + 3^{-1/2}\overline{h_{00}}(2L_{(+}L_{-)} + 2\delta^2 L_3^2)$$
$$+ 6^{1/2}(-4Ag^{-1}(\pi_A - \pi_B)\overline{p_{20}} + Ag^{-1}(2\pi_A + \pi_B)\overline{p_{00}})N';$$

$$((h, p), D\mathscr{I}_A^*N'^A) = \pm i2^{1/2}((\overline{p_{2\pm2}} + \pi_A\overline{h_{2\pm2}})L_{\mp} + (\overline{p_{2\pm1}} + \pi_A\overline{h_{2\pm1}})\delta^{-1}$$
$$- (\overline{p_{2\pm1}} + \pi_B\overline{h_{2\pm1}})\delta(\pm L_3 + n) - (\overline{X_p} + \tfrac{1}{2}\pi_B\,\overline{Y_h})L_{\pm})N'^{\pm1}$$
$$+ i(\delta^{-1}(\overline{p_{2\pm1}} + \pi_A\overline{h_{2\pm1}})L_{\mp} - (2\overline{Y_p} + \pi_B\overline{Y_h} + 2\pi_A\overline{X_h})L_3)N'^0. \quad \text{(IV.10)}$$

We have used repeated $\pm$ symbols to indicate a sum over upper and lower signs in sign-independent expressions like $\mathscr{H}'$ and $\mathscr{J}'_0$ only:

$$|h_{2\pm2}|^2 = \overline{h_{2\pm2}}h_{2\pm2} = |h_{2+2}|^2 + |h_{2-2}|^2,$$

$$\pm L_\pm h_{2\pm1} = L_+ h_{2+1} - L_- h_{2-1}. \tag{IV.11}$$

Brackets in the Hamiltonian enclose pairs of terms that are equal upon integration by parts. Poisson brackets of p_{sm} with either term in a pair agree and are easily evaluated by considering the term containing $\bar{h}_{sm}$. We have also introduced the abbreviations X_h, Y_h, and X_p, Y_p for the following combinations of the $m = 0$ spherical components:

$$X_h = 6^{-1/2}(h_{20} + \sqrt{2}h_{00}), \qquad Y_h = 6^{-1/2}(2h_{20} - \sqrt{2}h_{00}). \tag{IV.12}$$

With the following choice of the linearized lapse used by Bonanos (who also assumes $N = 1$)

$$N' = -\tfrac{1}{2}Nh_3^3 = -\tfrac{1}{2}NY_h, \tag{IV.13}$$

one can obtain equations satisfied by the "zero ξ-spin variables" and from these, the vacuum limit of the second-order equations for the variables used by Bonanos in the perfect fluid case. For $m \neq 0$ these variables are closely related to the real and imaginary parts of the "zero ξ-spin variables." In fact by keeping track of the background constraint function $\mathscr{H}$ and the Einstein components ${}^4G^a{}_b$ in the evaluation of the linearized Hamiltonian and Hamiltonian equations and with the proper insertion of the perfect fluid driving terms into the homogeneous vacuum equations, one can recover the perfect fluid perturbation equations themselves.

V. Harmonic Analysis

The action of either left or inverse right translation of $L^2(G)$ by dragging along is an infinite-dimensional unitary representation of G (the "left regular representation" or the "right regular representation"), which may be decomposed into a direct sum/integral of irreducible unitary representations for the groups of interest to us [22]. An orthonormal basis $\{Q^j\}$ of $L^2(G)$ adapted to each of these decompositions consists of the matrix elements of certain of the irreducible unitary representations of G. [Whenever j involves continuous indices, the corresponding elements of this basis do not themselves belong to $L^2(G)$ but satisfy a delta function normalization as in (III.15) and hence must be integrated against square-integrable functions of the continuous indices to yield elements of $L^2(G)$.]

Any element f in $L^2(G)$ then has a generalized expansion

$$f = \sum_j f^j Q^j, \qquad f^j = \langle Q^j, f \rangle, \tag{V.1}$$

where $\sum_j$ is a sum/integral over an appropriate index space in which j takes values. The collection of indices symbolized by j must include one index or a pair of indices specifying the irreducible unitary representation and two indices labeling the matrix elements for that given representation. In the nondegenerate case for example, a parameter associated with the image of the Casimir operator L^2 (together with an additional parameter for types VII_0 and VIII [23]) is necessary to specify the representation, while if the image of the operator L_3 in each representation is diagonal, the matrix element indices may be assumed to take values in the corresponding set of eigenvalues of that image operator in each representation in the direct sum/integral. The basis then consists of simultaneous eigenfunctions of $L^2 = \tilde{L}^2$, $\tilde{L}_3$, and L_3 and the shift operators $\tilde{L}_\pm$ and $L_\pm$ raise and lower the eigenvalues of the latter two operators; symbolically we may write:

$$L_\pm Q^j = \varepsilon_\pm^j Q^{j \pm 1}. \tag{V.2}$$

For Bianchi type II, L_3 is itself a Casimir operator and a slightly different scheme is used; one seeks simultaneous eigenfunctions of L_3, $\tilde{L}_1$, and $L_{(+}L_{-)}$, which have the interpretation of "transition matrix elements" rather than matrix elements with respect to a fixed basis. For Bianchi type I with the R^3 topology, ordinary Fourier integration is used and the basis consists of simultaneous eigenfunctions of $\{L_a\}$.

In each case, the functions $\{Q^j\}$ are eigenfunctions of $L_{(+}L_{-)}$ and L_3 with real eigenvalues. [For type VII_0, $L_{(+}L_{-)} = L^2$, while for types VIII and IX, $L_{(+}L_{-)} = \mathcal{N}(L^2 - L_3^2)$.] Except in the abelian case (and some abelianlike representations for type II) in which $L_\pm$ are diagonal, the operators $L_\pm$ act as shift operators on these eigenvalues and the quantities $\varepsilon_\pm^j$ may be assumed to be real. Since we are interested in real functions, it is worth noting that the eigenfunctions Q^j satisfy the following reality conditions that are inherited by the expansion coefficients f^j of a real function $f \in L^2(G)$:

$$\overline{Q^j} = (-1)^{\sigma(j)} Q^{\bar{j}}, \tag{V.3}$$

where $\bar{j}$ are the indices associated with the complex conjugate representation. The relevant spectra, the values of $\varepsilon_\pm^j$ and $\sigma(j)$ and other details of the harmonic analysis sketched here may be found in the author's dissertation [11].

The tensor product of the basis $\{Q^j\}$ with a given spherical basis $\{e_A\}$, $\{\omega^A\}$, $\{W^{sm}\}$, $\{\hat{W}^{sm}\}$, or $\{E_{sm}\}$ provides an orthogonal basis of the space of square-integrable tensor fields of that type; the expansion coefficients in this basis are just the expansion coefficients of the spherical components in the basis $\{Q^j\}$:

$$X^j_A = Q^j\omega^A, \qquad \hat{X}^j_A = Q^j e_A;$$
$$T^j_{sm} = Q^j W^{sm}, \qquad \hat{T}^j_{sm} = Q^j \hat{W}^{sm}, \qquad T^{\#j}_{sm} = Q^j E_{sm}. \tag{V.4}$$

The inner product of two complex-valued $\binom{p}{q}$ tensor fields S and T is given by

$$\langle S, T \rangle = \int_G \omega^{123} g^{1/2} \overline{S^{a\cdots}}_{b\cdots} T_{a\cdots}{}^{b\cdots}$$
$$= g^{1/2} \langle \overline{S^{a\cdots}}_{b\cdots}, T_{a\cdots}{}^{b\cdots} \rangle, \tag{V.5}$$

where the indices on T are raised and lowered with respect to $\mathscr{J}$. $\{g^{-1/4} T^j_{sm}\}$ is an orthonormal basis, for example. Each such basis satisfies reality conditions of the following kind that are inherited by the expansion coefficients of real tensor fields:

$$\overline{T^j_{sm}} = (-1)^{\sigma(j)+m} T^{\bar{j}}_{s-m}. \tag{V.6}$$

For tensor densities, the factor $g^{1/2}$ is replaced by $g^{-1/2}$ in (V.5).

In the nondegenerate case, the spherical bases are eigenvectors of both L_3 and $M_3 = L_3 + \tilde{L}_3$ with coinciding eigenvalues and the tensor product bases consist of simultaneous eigenvectors of $\tilde{L}^2$, $\tilde{L}_3$, L_3, and M_3 (but not L^2). The elements of these bases can be grouped into bases of each space of fixed eigenvalues of $\tilde{L}^2$, $\tilde{L}_3$, and L_3 (and hence of M_3 as well); the generic dimension of these spaces equals the number of independent components of the type of tensor field involved. For example, in the type IX case j stands for JMN and Q^{JMN} is an eigenfunction of $\tilde{L}^2 = L^2$, L_3, $\tilde{L}_3$, and M_3 with eigenvalues $J(J+1)$, M, N, and $M+N$, while A^{JMN}_A and T^{JMN}_{sm} are eigenvectors of $\tilde{L}^2$, $\tilde{L}_3$, L_3, and M_3 with respective sets of eigenvalues $J(J+1)$, M, $\mathscr{M} = N + A$, $\mathscr{M} + M$ and $J(J+1)$, M, $\mathscr{M} = N + m$, $\mathscr{M} + M$. $\tilde{L}_\pm$ act as raising and lowering operators on the eigenvalue M while $L_\pm$ act as such on the eigenvalue of N of Q^{JMN} only:

$$\tilde{L}_\pm Q^{JMN} = \varepsilon^{JM}_\pm Q^{JM\pm 1N}, \qquad L_\pm Q^{JMN} = \varepsilon^{JN}_\pm Q^{JMN\pm 1};$$
$$\tilde{L}_\pm T^{JMN}_{sn} = \varepsilon^{JM}_\pm T^{JM\pm 1N}_{sm}, \qquad \varepsilon^{JM}_\pm = (J(J+1) - M(M \pm 1))^{1/2}. \tag{V.7}$$

The spaces $F^{JM\mathscr{M}}$, $X^{JM\mathscr{M}}$, and $T^{JM\mathscr{M}}$ of functions, one-forms and covariant tensor fields of fixed eigenvalues of $\tilde{L}^2$, $\tilde{L}_3$, and L_3 in general have dimensions 1, 3, and 6 with bases:

$$\{Q^{JM\mathscr{M}}\} \qquad \{X^{JMN}_A | N + A = \mathscr{M}\} \qquad \{T^{JMN}_{sm} | N + m = \mathscr{M}; s = 0, 2\}. \tag{V.8}$$

Upper and lower bounds on the eigenvalue N associated with nontrivial kernels of the operators $L_\pm$ on $L^2(G)$ reduce these dimensions. For example, the spaces $T^{JM \pm (J + m_0)}$ have dimensions 1, 2, 4, 5, 6, for $m_0 = 2, 1, 0, -1, -2$, and only those spherical components with $\pm m \geq m_0$ are nonzero. The spaces $X^{JM \pm (J + m_0)}$ and $F^{JM \pm (J + m_0)}$ have dimensions 0, 1, 2, 3, 3, and 0, 0, 1, 1, 1 for that range of m_0, and only those spherical components with $\pm A \geq m_0$ are nonzero in the first case. For types VIII and VII$_0$, the single index J associated with $\tilde{L}^2$ must be supplemented by an additional parameter. In all three cases let I symbolize the representation parameters and the eigenvalues of $\tilde{L}_3$ and L_3 ($JM\mathcal{M}$ in the type IX case). The spaces F^I, X^I, T^I, etc., are important since LRS spatially homogeneous tensor equations couple only the spaces with common index I, since they are the simultaneous eigenspaces of a maximal set of commuting operators constructed from the Killing Lie derivatives that commute with all LRS spatially homogeneous linear differential operators. For Bianchi type II the corresponding spaces are not as easily described in terms of the eigenvalues; since $\{Q^j\}$ are here eigenfunctions of L_3, $L_{(+}L_{-)}$, and $\tilde{L}_1$, the tensor product bases consist of eigenvectors of L_3, $\tilde{L}_1$, and $L_{(+}^{\mathrm{orb}}L_{-)}^{\mathrm{orb}}$, where the orbital operators L_a^{orb} are defined as in (III.3). For type I the tensor product bases consist of eigenvectors of $L_a = L_a^{\mathrm{orb}}$ and the corresponding spaces F^I, X^I, T^I, etc., are just the spaces of fixed Fourier coefficients of the spherical components.

The linearized Hamiltonian equations for the variables $(h, p; N', \mathbf{N}')$ reduce the ODEs on each space

$$S^I = T^I \oplus (*T^\sharp)^I \oplus F^I \oplus \hat{X}^I, \qquad (\text{V.9})$$

where $(*T^\sharp)^I$ is the space of tensor densities corresponding to $T^{\sharp I}$. To evaluate the coefficients of such equations with respect to the appropriate eigentensor bases, using type IX as an example, one may simply replace $(h_{sm}, p_{sm}; N', N'^A)$ by $(h_{sm}^{JM\mathcal{M}-m}, p_{sm}^{JM\mathcal{M}-m}; N'^{JM\mathcal{M}}, \mathbf{N}_A^{JM\mathcal{M}-A})$ and the operators $L_3, L_\pm, L_{(+}L_{-)}$ by the appropriate constants, i.e.:

$$L_+^2 h_{sm} \to \varepsilon_+^{J\mathcal{M}-m+1}\varepsilon_+^{J\mathcal{M}-m}h_{sm}^{JM\mathcal{M}-m};$$
$$L_3 \to N, \qquad L_{(+}L_{-)} \to J(J+1) - N^2. \qquad (\text{V.10})$$

Because of the orthonormality of the basis $\{Q^j\}$, one may similarly evaluate the contribution to the summand/integral of a real quadratic functional like the Hamiltonian or the conserved quantities from S^I or $P^I = T^\oplus (*T^\sharp)^I$, respectively. However, it must always be accompanied by the corresponding contribution from the complex conjugate space because of the reality conditions. Canonically conjugate variables h_{sm}^j and $\overline{p_{sm}^j} = (-1)^{\sigma(j)+m}p_{s-m}^j$ belong to complex conjugate representations, so it is only the sum of the two contributions that generates the correct Hamiltonian equations for the canonical variables in P^I. (For type IX, $\bar{j}$ stands for $J - M - N$.) Similarly, only the sum of the contributions to the real conserved quantities is real and a constant of the motion for the finite dimensional dynamics on S^I. For example:

$$
\begin{aligned}
E_3^{JM\mathcal{M}} + E_3^{JM-\mathcal{M}} &= \mathcal{M}(i\langle p_{sm}^{JM\mathcal{M}-m}, h_{sm}^{JM\mathcal{M}-m}\rangle - i\langle p_{sm}^{J-M-\mathcal{M}-m}, h_{sm}^{J-M-\mathcal{M}-m}\rangle) \\
&= -2\mathcal{M}\,Im\langle p_{sm}^{JM\mathcal{M}-m}, h_{sm}^{JM\mathcal{M}-m}\rangle \\
&= 2\mathcal{M}\,Im\langle p_{sm}^{J-M-\mathcal{M}-m}, h_{sm}^{J-M-\mathcal{M}-m}\rangle.
\end{aligned} \tag{V.11}
$$

VI. The Moncrief Decomposition

The vacuum gravitational configuration space for spacetimes with manifold $R \times G$ is $\mathcal{M} = \text{RIEM}(G)$, the space of Riemannian metrics on G. The corresponding phase space is $P = T^*\mathcal{M} \sim \mathcal{M} \times S_d^2$ (the tilde indicates a natural identification), where $S_2 \supset \mathcal{M}$ is the space of symmetric $\binom{0}{2}$ tensor fields on G, and S_d^2 is the dual space of symmetric $\binom{2}{0}$ tensor densities on G. The natural symplectic two-form Ω and metric $\mathcal{G}$ on P are defined by:

$$\Omega((h_1, p_1), (h_2, p_2)) = \langle p_2, h_1 \rangle - \langle p_1, h_2 \rangle;$$

$$\mathcal{G}((h_1, p_1), (h_1, p_2)) = \langle h_1, h_2 \rangle + \langle p_1, p_2 \rangle;$$

$$\langle h_1, h_2 \rangle = \int_G \omega^{123} g^{1/2} \, \text{Tr} \, \hat{h}_1 \llcorner \hat{h}_2, \qquad h_i \in S_2; \tag{VI.1}$$

$$\langle p_1, p_2 \rangle = \int_G \omega^{123} g^{-1/2} \, \text{Tr} \, \hat{p}_1 \llcorner \hat{p}_2, \qquad p_i \in S_d^2.$$

These are compatible in the sense that the contravariant form of Ω with respect to $\mathcal{G}$ coincides with the symplectic inverse Ω^{-1} of Ω; let $\hat{\Omega} = "\mathcal{G}^{-1} \, \llcorner \, \Omega"$ be the mixed form of Ω with respect to G:

$$\hat{\Omega}((p_1, h_1), (h_2, p_2)) = \Omega(*^{-1}(h_1, p_1), (h_2, p_2)). \tag{VI.2}$$

Associated with the tensor fields Ω^{-1} and $\mathcal{G}^{-1}$ on P are linear maps from each cotangent space $T^*P_{(g,\pi)} \sim S_d^2 \times S_2$ onto the tangent space $TP_{(g,\pi)} \sim S_2 \times S_d^2$, which we shall denote by J and $*$, with inverses J^{-1} and $*^{-1}$. Let $\mathbb{J}: S_2 \times S_*^2 \to S_2 \times S_*^2$ be the linear transformation of each tangent space into itself associated with $\hat{\Omega}$:

$$\mathbb{J} = J \circ *^{-1}, \qquad \mathbb{J}^2 = -1, \qquad -\mathbb{J} \circ J = *;$$

$$\mathbb{J}(h, p) = (g^{-1/2}p^\flat, -g^{1/2}h^\sharp), \tag{VI.3}$$

$$J(p, h) = (h, -p), \qquad *(p, h) = (g^{-1/2}p^\flat, g^{1/2}h^\sharp).$$

We follow the notational conventions of Fischer and Marsden [5]; $\#$ and $\flat$ indicate the covariant and contravariant forms of contravariant and covariant tensors or tensor densities, respectively. In terms of spherical components, (IV.1) and (VI.3) take the following form:

$$\Omega((h_1, p_1), (h_2, p_2)) = \langle p_{2sm}, h_{1sm} \rangle - \langle p_{1sm}, h_{2sm} \rangle$$

$$\mathscr{G}((h_1, p_1), (h_2, p_2)) = g^{1/2}\langle h_{1sm}, h_{2sm} \rangle + g^{-1/2}\langle p_{1sm}, p_{2sm} \rangle;$$

$$\mathbb{J}(h_{sm}, p_{sm}) = (g^{-1/2}p_{sm}, -g^{1/2}h_{sm}); \tag{VI.4}$$

$$J(p_{sm}, h_{sm}) = (h_{sm}, -p_{sm}).$$

$\mathbb{J}$ is the natural almost complex structure associated with a compatible metric and symplectic two-form; it satisfies the following identities:

$$\mathscr{G}(\mathbb{J}X, \mathbb{J}Y) = \mathscr{G}(X, Y) = \Omega(X, -\mathbb{J}Y), \qquad X, Y \in S_2 \times S_d^2. \tag{VI.5}$$

The Moncrief decomposition is an orthogonal decomposition (orthogonal with respect to $\mathscr{G}$) of each tangent space $TP_{(g, \pi)} \sim S_2 \times S_d^2$ to a point $(g, \pi) \in C \subset P$ of the constraint space C:

$$C = \{(g, \pi) \in P \mid \phi(g, \pi) = 0\}, \qquad \phi = (\mathscr{H}, \mathscr{J}). \tag{VI.6}$$

This orthogonal direct sum has the following form [5,6]:

$$S_2 \times S_d^2 = (\ker D\phi \cap \ker D\phi \circ \mathbb{J}) \oplus \operatorname{Ran} J \circ D\phi^* \oplus \operatorname{Ran} - \mathbb{J} \circ J \circ D\phi^*$$

$$= \operatorname{GAUGE}_{(g, \pi)}^\perp \oplus \operatorname{GAUGE}_{(g, \pi)} \oplus \operatorname{CONSTRAINT}_{(g, \pi)}. \tag{VI.7}$$

The first subspace is invariant under $\mathbb{J}$ while $\mathbb{J}$ is an isomorphism between the second two subspaces. We are only interested in this decomposition at LRS left-invariant points of P and the tensor harmonic expansion technique we are using forces us to limit our attention to the L^2-tangent space to P at these points, i.e., the square-integrable subspace of $S_2 \times S_d^2$.

Since the operators involved in the Moncrief decomposition are LRS left-invariant linear differential operators on G, the decomposition projects to an orthogonal direct sum of each eigenspace P^I of the orthogonal direct sum/integral decomposition of the square-integrable subspace of $S_2 \times S_d^2$ (more precisely, of its complexification). On each space P^I, these operators are linear transformations (also involving F^I and $\hat{X}^I$) whose matrices with respect to the eigentensor bases are easily evaluated from (IV.10) and the remarks following (V.9). The components of these matrices are all real except for type I and the abelianlike type II representations.

The eigentensor basis of P^I is an orthogonal symplectic basis; in the type IX case it is explicitly

$$\{(T_{sm}^{JM\mathcal{M}-m}, 0), (0, {}^*T_{sm}^{\neq JM\mathcal{M}-m}) \mid s = 0, 2; m = 2, \ldots, -2\}. \tag{VI.8}$$

Normalizing this basis by factors of $g^{\pm 1/2}$ leads to an orthonormal symplectic basis of P^I, i.e., an "orthosymplectic" basis. By a symplectic basis we mean that it consists of conjugate pairs (E^j, F^j), which satisfy $\Omega(E^j, F^{j'}) = \delta_{jj'}$, although the ordering of the basis is left free. The expansion coefficients in a symplectic basis are canonical coordinates, with the (complex conjugate of the) F^j coordinate being the momentum conjugate to the E^j coordinate.

The Moncrief decomposition can be "solved" by finding a new orthosymplectic basis of P^I adapted to the orthogonal direct sum (VI.7). The linear transformation to this new basis is a canonical transformation with respect to the linearized canonical structure of TP, the linearized phase space. Since it involves $(\mathscr{g}, \pi)$ nontrivially, this is a time-dependent linear canonical transformation for each background solution of the full Hamiltonian equations. Moncrief [10a] has referred to this class of canonical transformations, although the most general such transformation mentioned there may be obtained by completing any basis of $\text{GAUGE}_{(\mathscr{g}, \pi)}$ to a symplectic basis of the whole tangent space.

. A basis for $(\text{GAUGE}^{\perp}_{(\mathscr{g}, \pi)})^I$ can be found by row reduction of the direct sum of the matrices of $(D\phi)^I$ and $(D\phi \circ \mathbb{J})^I$; the ordering of the orthogonal symplectic eigentensor basis in sequential conjugate pairs is crucial for this step. An orthosymplectic basis $\{E_i, -\mathbb{J}E_i\}$ can then be obtained by an obvious modification of the Gramm–Schmidt orthonormalization algorithm that incorporates the symplectic condition. $\{(J \circ D\mathscr{H}*1)^I, (J \circ D\mathscr{J}^*_A 1)^I\}$ is a basis of $(\text{GAUGE}_{(\mathscr{g}, \pi)})^I$ Applying Gramm–Schmidt to it yields an orthonormal basis $\{E_\alpha\}$ whose image by the map $-\mathbb{J}$ is an orthonormal basis of $(\text{CON-STRAINT}_{(\mathscr{g}, \pi)})^I$ such that the direct sum basis of the direct sum of the two subspaces is an orthosymplectic basis, as a consequence of (VI.5). Putting these bases together leads to an orthosymplectic basis of the whole space P^I adapted to the Moncrief decomposition:

$$\{E_{(R)}\} = \{E_i, -\mathbb{J}E_i; E_\alpha, -\mathbb{J}E_\alpha\}. \tag{VI.9}$$

Let $\{Z^R\} = \{h^j_{sm}, p^j_{sm}\}$ be the original canonical coordinates on P^I and $\{Z^{(R)}\} = \{Q^i, P_i; G^\alpha, C_\alpha\}$ be the new canonical coordinates:

$$Z^R = E^R_{(S)} Z^{(S)}. \tag{VI.10}$$

The Hamiltonian is a sum/integral of contributions from each space S^I consisting of a Hermitian quadratic form in the canonical coordinates of P^I and another Hermitian quadratic form involving cross terms between P^I and $F^I \times \hat{X}^I$. Because (VI.10) is a time-dependent linear canonical transformation, in addition to transforming the quadratic forms to the new canonical variables, we must add the term

$$(\Delta\text{HAM}^{\text{spherical}})^I = \tfrac{1}{2}\overline{Z^{(R)}E^U_{(R)}} J_{UV} \dot{E}^V_{(S)} Z^{(S)} \tag{VI.11}$$

This differs from (III.21) only in that it is a Hermitian rather than a symmetric quadratic form.

The contribution to the summand/integrand of the new Hamiltonian from both S^I and $\overline{S}^I$ generates the Hamiltonian equations of motion for the new canonical variables $\{Z^{(R)}\}$. The canonically conjugate variables (Q^i, P_i) of the subspace $(\text{GAUGE}^{\perp}_{(g, \pi)})^I$ are gauge invariant and decouple from the pure gauge variables G^{α} and the canonically conjugate linearized constraint variables C_{α}, which are also gauge invariant. ("Canonically conjugate" involves complex conjugation as well.) This decoupling occurs since the cross terms between (Q^i, P_i) and $\{G^{\alpha}\}$ in the Hermitian form Hamiltonian vanish. They must vanish in order that the linearized constraints $C_{\alpha} = 0$ be preserved (i.e., $\dot{C}_{\alpha} = 0$ when $C_{\alpha} = 0$). The reduced Hamiltonian for the gauge invariant variables is just that part of the Hamiltonian involving only (Q^i, P_i).

In the generic case when the (complex) dimension of P^I is 12, the linear algebra we have described for solving the Moncrief decomposition and transforming the Hamiltonian is too complicated to carry out by hand but could be programmed in a symbolic manipulation computer language. However, for types II, VIII, and IX, lower-dimensional spaces associated with the kernels of $L_{\pm}$ exist on which the linear algebra is manageable. The spaces for types II and VIII are exactly analogous to those for type IX described after (V.8). The spaces P^I for $m_0 = 2$ are automatically gauge invariant, but for $m_0 = 1$, $(\text{GAUGE}^{\perp}_{(g, \pi)})^I$ and $(\text{GAUGE}_{(g, \pi)})^I$ have dimension 2 and 1, respectively, and the adapted orthosymplectic basis is easily obtained. The transformation of the Hamiltonian is rather complicated and is under present investigation. The results of this special case should indicate whether gauge invariant perturbation theory is a simplification or complication in spatially homogeneous cosmology.

References

[1] Taub, A. H., *Ann. Math.* **53**, 472 (1951).
[2] Misner, C. W., and Taub, A. H., *Sov. Phys.—JETP* **28**, 122 (1969).
[3] Taub, A. H., and MacCallum, M. A. H., *Commun. Math. Phys.* **25**, 173 (1972).
[4] Jantzen, R. T., *Commun. Math. Phys.* **64**, 211 (1979); *in* "Relativistic Cosmology and Blanchi Universes" (R. Ruffini, ed.), to appear.
[5] Fischer, A. E., Marsden, J. E., *J. Math. Phys.* **13**, 546 (1972); *Proc. Int. Sch. Phys.* "Enrico Fermi" (1976).
[6] Moncrief, V., *J. Math. Phys.* **16**, 493, 1556 (1975); **17**, 1893 (1976).
[7] Smarr, L., and York, J. W., Jr., *Phys. Rev. D* **17**, 2529 (1978).
[8] York, J. W., Jr., *in* "Sources of Gravitational Radiation" (L. Smarr, ed.). Cambridge Univ. Press, London and New York, 1979.
[9] Taub, A. H., *Commun. Math. Phys.* **15**, 235 (1969); *in* "Lectures at the Centro Internazionale Matematico Estivo, Bressanone, 1970" (C. Cattaneo, ed.). Edizioni Cremonese, Rome, 1971.

[10] Moncrief, V., *Ann. Phys.* (*N. Y.*) **88**, 323, 343 (1974).

[10a] Moncrief, V., *Phys. Rev. D* **12**, 1526 (1975).

[11] Jantzen, R. T., Ph.D. Thesis, Univ. of California, Berkeley.

[12] Perko, T. E., Matzner, R. A., and Shepley, L. C., *Phys. Rev. D* **6**, 969 (1972).

[13] Hu, B. L., *J. Math. Phys.* **15**, 1748 (1974).

[14] Jantzen, R. T., *J. Math. Phys.* **19**, 1163 (1978).

[15] Lifshitz, E. M., and Khalatnikov, I. M., *Adv. Phys.* **12**, 185 (1963); Lifshitz, E. M., *J. Phys. USSR* **10**, 116 (1946).

[16] Gerlach, U. H., and Sengupta, U. K., *Phys. Rev. D* **18**, 1773 (1978).

[17] Collins, C. B., *J. Math. Phys.* **18**, 2116 (1977).

[18] Miller, J. G., *Commun. Math. Phys.* **52**, 1 (1977).

[19] Siklos, S. T. C., *Phys. Lett. A* **59**, 173 (1976).

[20] Bonanos, S., *Commun. Math. Phys.* **22**, 190 (1971); **26**, 259 (1972).

[21] Misner, C. W., Thorne, K. S., and Wheeler, J. A., "Gravitation." Freeman, San Francisco, California, 1973.

[22] Vilenkin, N. J., "Special Functions and the Theory of Group Representations." Am. Math. Soc., Providence, Rhode Island, 1968.

[23] Wybourne, B. G., "Classical Groups for Physicists." Wiley, New York, 1974.

9

Locally Isotropic Space–Times with Nonnull Homogeneous Hypersurfaces

M. A. H. MacCallum

Department of Applied Mathematics
Queen Mary College
London, England

Abstract

Using a technique developed by Schmidt, all possible Lie algebras of groups G_4 transitive on nonnull hypersurfaces in space–time are calculated. The actions of all subgroups G_3 of the G_4 are discussed, and the Ricci and Weyl tensors of the corresponding metrics are calculated and classified. The results are compared with those of other authors.

I. Introduction

It is a pleasure to dedicate this paper to Prof. A. H. Taub, not only in memory of many enjoyable and educative contacts with him (and more to come, I hope), but also because the subject is closely related to that of one of his most famous papers [1], which could indeed be said to have been the stimulus for all the subsequent work on groups of motions in space–time.

The subject of this paper is not a new one: Space–times with isotropy have been extensively considered in the literature. My aim here is to recover many known results by a different and, in my opinion, more elegant technique, and to provide a complete list of metrics, with Weyl and Ricci tensors, together with a full account of the action of all the subgroups G_3 of the G_4.

A space–time is said to be *locally isotropic* if, for every point p, there is a nontrivial group of motions, the isotropy group of p, under which p is fixed.

121

(The theory of Lie groups of transformations required here is described in detail in Eisenhart's classic text [2]; some recent accounts in more modern notation may be found in MacCallum [3,5] and Ryan and Shepley [4].) This definition has been shown to be equivalent to the others used in the literature [6–10].

Local isotropy arises when a group of motions acts multiply transitively on (sub)manifolds of space–time. If V_d denotes a Riemannian manifold of dimension d, and N_r a null submanifold of dimension r, the possibilities within a space–time are a G_3 on V_2 or N_2, a G_r on V_3 or N_3 ($4 \le r \le 6$), and a G_r on V_4 ($r \ge 5$). Petrov [11] and Defrise [9] have described methods for finding all locally isotropic spacetimes and given lists of the resulting metrics. All spacetimes with local rotational symmetry (i.e., with a spatial rotation as the isotropy) were found by Ellis and Stewart [6,7].

In this paper I consider only the case of a G_4 on nonnull orbits V_3, but the method used can be applied to other cases with nonnull orbits. (The construction starts by working on a single orbit, and thus does not readily extend to the cases with null orbits; a single null orbit may be the boundary between two regions where the orbits are nonnull, or where the group is simply transitive.) Schmidt [12] applied the method outlined here to the case of G_r on V_4 ($r \ge 5$).

Schmidt's procedure is described in Section II, and applied to the case in hand in Sections III–V. These sections give the possible G_4, their subgroups G_3, and the line elements for the distinct metrics. Section VI gives the Weyl and Ricci tensors and their classification, and Section VII concludes by relating the results of the present work to some of the many known results in the literature.

In giving the metrics, the signature is taken to be $+2$ and the signs of the Ricci and Weyl tensors are defined by the conventions

$$u_{a;bc} - u_{a;cb} = R^d{}_{abc} u_d \tag{I.1}$$

for arbitrary u_a, and

$$R_{ab} = R^d{}_{adb}, \qquad R = R^a{}_a,$$

$$C^{ab}{}_{cd} = R^{ab}{}_{cd} + 2g^{[a}{}_{[d} R^{b]}{}_{c]} + \frac{R}{3} g^{[a}{}_{[c} g^{b]}{}_{d]}, \tag{I.2}$$

where the usual index conventions apply (i.e., Latin indices range from 1 to 4, x^4 being the time coordinate, repeated indices are summed over, and indices inside square brackets are skewed over, and the semicolon denotes covariant differentiation with respect to the basis vectors corresponding to the following indices). The computations whose results are reported in Section VI were carried out using differential forms, taking either an orthonormal [6,7] or a complex null [14] basis tetrad (for some details, see, e.g. [4]).

The algebraic classification of the Weyl tensor used in Section VI is due to Petrov [11]. The existence of an isotropy shows immediately that the Weyl tensor is of Petrov type D (in the case of spatial rotation or boost symmetry) or type N (in the case of null rotation symmetry) or zero (cp. [15]). The Ricci tensor is classified in Section VI by its Segré characteristic, which gives the geometric multiplicities of the eigenvalues; equal eigenvalues are enclosed in a round bracket, complex eigenvalues are denoted Z, and numbers before the comma refer to spacelike eigenvectors. A useful exposition of this classification was given by Hall [16]. Local rotational symmetry permits only types $[(11)1,1]$, $[(11),Z\bar{Z}]$, and $[(11),2]$ and their specializations, while local boost symmetry (i.e., where the isotropy is a proper Lorentz transformation, or boost) permits only type $[11(1,1)]$ and its specializations, and local null rotation symmetry (defined in an obvious way) permits types $[1(11,1)]$, $[1(1,2)]$, and $[(1,3)]$ and specializations. (It may be helpful to remind the reader that the cosmological constant is type $[(111,1)]$; a perfect fluid, type $[(111),1]$; tachyonic fluid, type $[1(11,1)]$; nonnull electromagnetic fields, type $[(11)(1,1)]$; and pure radiation fields, including null electromagnetic fields, type $[(11,2)]$.)

II. Description of the Calculation Technique

Schmidt [12,13] has shown how to find all multiply transitive Lie algebras of Killing vectors on Riemannian spaces. The method can be expressed as the following algorithm.

(i) *Specify the dimension d and signature of the Riemannian space V_d.* In our case $d = 3$ and the orbits may be timelike (signature $+1$) or spacelike (signature $+3$). The cases with timelike orbits must be static or stationary.

(ii) *Specify the dimension r of the Lie algebra ($r > d$).* In our case $r = 4$. Each point then has an isotropy group of dimension $r - d$ (in our case, 1). The isotropy groups of different points are conjugate subgroups of the full group of motions, so we may speak of *the* isotropy group. (Although our calculation concerns only the Lie algebra, Schmidt [13] has shown that a corresponding group essentially exists, in the sense that any small neighborhood in V_d is isometric to a neighborhood in a space on which such a group acts.) The isotropy group of a point p is isomorphic to the linear isotropy group, that it induces on the tangent space to V_d at p, and must therefore be a subgroup of the appropriate generalized orthogonal group. In our V_3, the latter is the rotation group SO(3) for spacelike V_3 and the three-dimensional Lorentz group SO(2, 1) for timelike V_3.

(iii) *Choose a point p, and specify the linear isotropy group, choosing a suitable basis $\{Y_\alpha\}$ of its Lie algebra.* (We shall, with some loss of rigor, use the same notation for the isotropy group generators.) For our problem, the possibilities are

Case (A): local rotational symmetry; the isotropy is a spatial rotation and the V_3 may be spacelike or timelike.

Case (B): local boost symmetry; the isotropy is a boost and the V_3 is timelike.

Case (C): local null rotation symmetry; the isotropy is a null rotation and the V_3 is timelike.

The isotropy group of p is generated by a Lie algebra of Killing vector fields with basis $\{Y_\alpha; \alpha = 1, 2, \ldots, s\}$ such that at p, $Y_\alpha = 0$. The Y_α can be chosen to give a standard form for the commutators $[Y_\alpha, Y_\beta]$. (In our case this is trivial.)

(iv) *Take Killing vectors $\{X_i; i = 1, \ldots, d\}$ such that $X_i \neq 0$ at p and the X_i and Y_α together form a basis of the whole algebra.* At p the X_i form a basis of the tangent space, and this can be chosen so that the values of the X_i at p are adapted to the action of the Y_α on this space. The commutators $[X_i, Y_\alpha]$ are then known at p from this action. In our problem we write

$$[Y, X_1] = DY,$$
$$[Y, X_2] = CX_1 + BX_2 + AX_3,$$
$$[Y, X_3] = (C - A)X_2 + BX_3. \tag{II.1}$$

A, B, and C are zero, except in cases (A), (B), and (C), respectively, where they are 1. In case (A), X_1 is a spacelike or timelike unit vector at p perpendicular to the plane of rotation, and X_2 and X_3 are spacelike unit vectors at p in the plane of rotation. Possible terms in Y in $[Y, X_2]$ and $[Y, X_3]$ have been eliminated by adding multiples of Y to the original choices of $X_1, X_2,$ and X_3; the same has been done in cases (B) and (C). In case (B), X_2 and X_3 are null vectors, with $X_2 \cdot X_3 = -1$, in the plane of the boost, and X_1 is an orthogonal unit spacelike vector, at p. In case (C), X_1 and X_3 are null vectors, with $X_1 \cdot X_3 = -1$, and X_2 is an orthogonal spacelike unit vector, at p; X_1 and X_2 lie in the plane of the null rotation. The remaining freedom of basis in (II.1) is $X_1' = X_1 + \alpha Y$ in cases (A) and (B), and $X_3' = X_3 + \gamma Y$ in case (C), where α and γ are constants.

(v) *Impose the Jacobi identities.* This determines the possible values of the remaining unknown structure constants, i.e., $[X_i, X_j]$, and the coefficient of Y_β in $[Y_\alpha, X_i]$. We write the remaining commutators for the present problem as

$$[X_2, X_3] = N_1 X_1 + (N_{12} - M_3)X_2 + (N_{13} + M_2)X_3 + L_1 Y,$$
$$[X_3, X_1] = (N_{12} + M_3)X_1 + N_2 X_2 + (N_{23} - M_1)X_3 + L_2 Y, \tag{II.2}$$
$$[X_1, X_2] = (N_{13} - M_2)X_1 + (N_{23} + M_1)X_2 + N_3 X_3 + L_3 Y.$$

We leave consideration of the restrictions imposed on (II.1) and (II.2) by the Jacobi identities to Sections III–V, which deal with cases (A), (B), and (C), respectively.

Schmidt has shown how to calculate the curvature of V_d at p directly from these commutators [13]. However, in our case the orbits are only submanifolds of the spacetime, and so, for finding the spacetime curvature, it is necessary to add some further steps to the computation.

By a well-known theorem [2], the normals to the group orbits are geodesic. The metric for spacelike orbits is thus

$$ds^2 = g_{ij}dx^i\, dx^j - dt^2, \tag{II.3}$$

where $i, j = 1, 2, 3$, while in the timelike case it is

$$ds^2 = dt^2 + g_{ij}\, dx^i\, dx^j, \tag{II.4}$$

where, in either case,

$$d\sigma^2 = g_{ij}dx^i\, dx^j \tag{II.5}$$

gives the metric on the orbits V_3. The coordinates x^i ($i = 1, 2, 3$) can be "comoving" with respect to the coordinate t. If the G_4 contains a subgroup G_3 simply transitive on the V_3, one can choose a triad $\{\mathbf{e}_a\}$ (of reciprocal group generators) invariant under this subgroup, and then

$$d\sigma^2 = g_{ab}\omega^a\omega^b, \tag{II.6}$$

where the ω^a are dual to the $\mathbf{e}_a$, and are t-independent, and the g_{ab} depend only on t. Coordinate forms for all types of G_3 have been given by several authors (e.g. [1,4,5]).

If there is a simply transitive G_3, it must have a basis of Killing vectors $\{\mathbf{Z}_i\}$ such that $\mathbf{Z}_i = \mathbf{X}_i$ at p. Thus

$$\mathbf{Z}_1 = \mathbf{X}_1 + \alpha\mathbf{Y}, \qquad \mathbf{Z}_2 = \mathbf{X}_2 + \beta\mathbf{Y}, \qquad \mathbf{Z}_3 = \mathbf{X}_3 + \gamma\mathbf{Y}. \tag{II.7}$$

(In a more general case, with $s > 1$, one would take $\mathbf{Z}_i = \mathbf{X}_i + A_i^\alpha\mathbf{Y}_\alpha$ with some constant matrix A_i^α.) The possible values of the unknown constants α, β, and γ are found by imposing the condition that the $\mathbf{Z}_i$ generate a Lie algebra (i.e., in our case, that the $[\mathbf{Z}_i, \mathbf{Z}_j]$ contain no term in $\mathbf{Y}$). Thus,

(vi) *Find all simply transitive subgroups G_3 by calculating the commutators of* (II.7). The form (II.6) can be further simplified by choosing the $\mathbf{e}_a$ at p so that there $\mathbf{e}_a = \mathbf{X}_a$ ($a = 1, 2, 3$). The vector field $\mathbf{n} = \partial/\partial t$ normal to the orbits must obey

$$0 = \mathscr{L}_\xi\mathbf{n} = [\xi, \mathbf{n}] = -\mathscr{L}_\mathbf{n}\xi, \tag{II.8}$$

for any Killing vector ξ, and hence

$$\mathbf{Y} = 0 \tag{II.9}$$

at all points on the integral curve of **n** through t, $\Gamma(t)$ say. Thus **Y** generates the isotropy at all points of $\Gamma(t)$. The commutators are unchanged, while all the Killing vectors obey (II.8). Thus in case (A), along $\Gamma(t)$, $\mathbf{X}_1$ is always the axis of rotation, while $\mathbf{X}_2$ and $\mathbf{X}_3$ always lie in the plane of rotation and must scale by the same length scale factor. Similarly, in case (B) $\mathbf{X}_2$ and $\mathbf{X}_3$ are null and scale by the same factor along $\Gamma(t)$, while $\mathbf{X}_1$ remains orthogonal to them, and in case (C) $\mathbf{X}_1$ is null and $\mathbf{X}_2$ is spacelike along $\Gamma(t)$ and they both scale by the same factor, but $\mathbf{X}_3$ may become spacelike or timelike, provided that its component along the null direction **V** in $t = $ constant such that $\mathbf{X}_1 \cdot \mathbf{V} = -1$ and $\mathbf{X}_2 \cdot \mathbf{V} = 0$ scales like $\mathbf{X}_1$ and $\mathbf{X}_2$, and $\mathbf{X}_3$ lies in the $(\mathbf{V}, \mathbf{X}_1)$ plane.

The arguments in the last paragraph show that one can choose a basis σ^a of t-independent one-forms dual to a basis of reciprocal group generators aligned with the $\mathbf{X}_i$ at all points of $\Gamma(t)$, if there is a simply transitive G_3. Moreover, the $d\sigma^a$ are known

$$d\sigma^a = C^a{}_{bc}\,\sigma^b \wedge \sigma^c, \tag{II.10}$$

where $C^a{}_{bc}$ are the structure constants of the simply transitive group, which may be scaled to a canonical form by scaling the σ^a. The metrics are

$$ds^2 = e(-dt^2 + E^2\sigma^1\sigma^1) + F^2(\sigma^2\sigma^2 + \sigma^3\sigma^3) \tag{II.11}$$

for case (A), $e = 1$ giving spacelike, and $e = -1$ timelike, orbits;

$$ds^2 = dt^2 + E^2\sigma^1\sigma^1 - 2F^2\sigma^2\sigma^3, \tag{II.12}$$

in case (B); and

$$ds^2 = dt^2 + E^2[\sigma^2\sigma^2 - 2\sigma^3(\sigma^1 + F\sigma^3)], \tag{II.13}$$

in case (C), where in all cases E and F are functions of t alone.

Incidentally, it may be noted that the method described here gives the simply transitive subgroups without ambiguity, in that one cannot obtain two distinct bases of the same subgroup G_3. There may, however, be one more subgroup G_3, which must be multiply transitive. When no simply transitive subgroup exists, we can use a multiply transitive subgroup to help us in constructing the metric, using the well-known metrics for V_2 and N_2 invariant under a G_3, together with results such as those in Schmidt [17].

(vii) *Find any multiply transitive subgroup G_3.* Such a subgroup must have a basis $\{\mathbf{X}_2, \mathbf{X}_3, \mathbf{Y}\}$ in cases (A) and (B), or $\{\mathbf{X}_1, \mathbf{X}_2, \mathbf{Y}\}$ in case (C). It will thus be unique, and together with the groups found at step (vi), completes the list of subgroups G_3 of the G_4. For these bases to generate a subgroup we require $N_1 = 0$ [cases (A) and (B)] or $N_3 = 0$ [case (C)]. From (II.9), this subgroup will be multiply transitive at all points on $\Gamma(t)$, but this only ensures multiple transitivity in a three-dimensional manifold containing $\Gamma(t)$ and the two-dimensional orbits intersecting $\Gamma(t)$.

The additional conditions needed to make the subgroup multiply transitive throughout V_4 can be found as follows. If an isometry moves p to q, it must move the plane in which the isotropy acts at p to the similar plane at q. The multiply transitive subgroup in a two-surface through p must also be Lie transported, so we have only to compare the Lie transported subgroup with the one generated by the same Killing vectors we started with. (Note that a Killing vector field is not necessarily invariant under Lie transport by other Killing vectors.) We thus find that a multiply transitive subgroup exists in cases (A) and (B) if

$$N_1 = N_{12} + M_3 = N_{13} - M_2 = 0, \tag{II.14}$$

and in case (C) if

$$N_3 = N_{23} - M_1 = N_{13} + M_2 = 0. \tag{II.15}$$

III. Locally Rotationally Symmetric Space–Times

The LRS spaces compose case (A). Here the Jacobi identities for the triples $(\mathbf{X}_2, \mathbf{X}_3, \mathbf{Y})$, $(\mathbf{X}_3, \mathbf{X}_1, \mathbf{Y})$, and $(\mathbf{X}_1, \mathbf{X}_2, \mathbf{Y})$ yield†

$$N_{12} = N_{13} = N_{23} = M_2 = M_3 = L_2 = L_3 = D = 0, N_2 = N_3. \tag{III.1}$$

Using the basis freedom in $\mathbf{X}_1$, we can set $N_1 L_1 = 0$, and this, together with the remaining Jacobi identity [for $(\mathbf{X}_1, \mathbf{X}_2, \mathbf{X}_3)$], gives

$$M_1 N_1 = N_1 L_1 = L_1 M_1 = 0. \tag{III.2}$$

The conditions for (II.7) to generate a simply transitive subgroup are

$$L_1 - \alpha N_1 + \beta^2 + \gamma^2 = 0,$$
$$\beta(N_2 + \alpha) - M_1 \gamma = 0, \tag{III.3}$$
$$\beta M_1 + \gamma(N_2 + \alpha) = 0.$$

Thus there are two types of basis choice, namely,

$$\beta = \gamma = 0, \qquad N_1 \alpha = L_1, \tag{III.4a}$$
$$\alpha = -N_2, \qquad M_1 = 0, \qquad \beta^2 + \gamma^2 + L_1 = 0. \tag{III.4b}$$

---

† $D = 0$ can be derived from more sophisticated consideration of the weakly reductive character of the isotropy (see [12, 13]).

The commutators for the simply transitive group would be

$$[\mathbf{Z}_2, \mathbf{Z}_3] = N_1\mathbf{Z}_1 - \beta\mathbf{Z}_2 - \gamma\mathbf{Z}_3,$$
$$[\mathbf{Z}_3, \mathbf{Z}_1] = (N_2 + \alpha)\mathbf{Z}_2 - M_1\mathbf{Z}_3, \qquad \text{(III.5)}$$
$$[\mathbf{Z}_1, \mathbf{Z}_2] = M_1\mathbf{Z}_2 + (N_2 + \alpha)\mathbf{Z}_3.$$

Using condition (III.2) leads us to the following cases:

A1. $N_1 = M_1 = 0$

Equation (II.14) shows there is always a multiply transitive subgroup. Its Bianchi type is VII_0, VIII, or IX, respectively, when L_1 is zero, negative, and positive; Schmidt's method [13] easily shows that it acts on two surfaces of, respectively, zero, negative, and positive curvature, i.e., planes, pseudospheres, and spheres. Considering each case in turn we find

(a) $L_1 = 0$: in this case (III.4a) gives a one-parameter family of groups G_3 of Bianchi type VII_0, and a unique group of Bianchi type I, also given by (III.4b). The metrics are

$$ds^2 = e(-dt^2 + E^2\,dx^2) + F^2(dy^2 + dz^2). \qquad \text{(III.6)}$$

(b) $L_1 < 0$: (III.4a) has no solution, but (III.4b) gives a one-parameter family of subgroups of Bianchi type III. The metric is

$$ds^2 = e(-dt^2 + E^2\,dx^2) + F^2(d\theta^2 + \sinh^2\theta\,d\phi^2). \qquad \text{(III.7)}$$

(c) $L_1 > 0$: in this case there is no simply transitive subgroup. The metric is of the Kantowski–Sachs form [18] if $e = 1$, or static spherically symmetric if $e = -1$; it is

$$ds^2 = e(-dt^2 + E^2\,dx^2) + F^2(d\theta^2 + \sin^2\theta\,d\phi^2). \qquad \text{(III.8)}$$

The metrics (III.6)–(III.8) can jointly be written as

$$ds^2 = e(-dt^2 + E^2\,dx^2) + F^2(d\theta^2 + f^2(\theta)\,d\phi^2), \qquad \text{(III.9)}$$

where $f = \sin\theta, \theta, \sinh\theta$, respectively, when $L_1 > 0, L_1 = 0, L_1 < 0$.

A2. $L_1 = M_1 = 0 \neq N_1$

There are no multiply transitive subgroups [see (II.14)]. There are three cases to consider.

(a) $N_2 = 0$: there is a unique simply transitive subgroup of Bianchi type II; the metric can be written as (II.11) with

$$\sigma^1 = dx + y\,dz, \qquad \sigma^2 = dy, \qquad \sigma^3 = dz. \qquad \text{(III.10)}$$

(b) $N_1 N_2 > 0$:   there is a unique simply transitive subgroup of Bianchi type IX; the metric can thus be written as (II.11) with

$$\sigma^1 = dx + \cos y \, dz, \qquad \sigma^2 = \cos x \, dy + \sin y \sin x \, dz,$$
$$\sigma^3 = -\sin x \, dy + \sin y \cos x \, dz, \qquad\qquad\qquad\text{(III.11)}$$

i.e., as

$$ds^2 = e[-dt^2 + E^2(dx + \cos y \, dz)^2] + F^2(dy^2 + \sin^2 y \, dz^2). \quad \text{(III.12)}$$

(c) $N_1 N_2 < 0$:   there is a simply transitive subgroup of Bianchi type VIII, and, by (III.4b), a one-parameter family of groups of type III. The metric is given by (II.11) with

$$\sigma^1 = dx + \cosh y \, dz, \qquad \sigma^2 = \cos x \, dy + \sinh y \sin x \, dz,$$
$$\sigma^3 = -\sin x \, dy + \sinh y \cos x \, dz, \qquad\qquad\qquad\text{(III.13)}$$

$$ds^2 = e[-dt^2 + E^2(dx + \cosh y \, dz)^2] + F^2(dy^2 + \sinh^2 y \, dz^2). \quad \text{(III.14)}$$

The metrics (A2) include the well-known vacuum Taub–NUT metrics [1,19]. They can be combined with (III.9), for example, in the form

$$ds^2 = e\{-dt^2 + E^2[dx + lf(y)dz]^2\} + F^2\{dy^2 + [f'(y)]^2 \, dz^2\}, \quad \text{(III.15)}$$

where $l = 0$ [case (A1)] or $l \neq 0$ [case (A2)], and $f(y)$ is $\cos y$, y, or $\cosh y$, respectively, when $k = 1, 0$, or -1, k being the sign of $N_1 N_2 + L_1$.

A3. $L_1 = N_1 = 0 \neq M_1$

Equation (II.14) shows there is a multiply transitive subgroup of Bianchi type VII$_0$, which acts on planes. There is a simply transitive subgroup of Bianchi type V, and, by (III.4a), a one-parameter family of groups of type VII$_h$. The metric has the form (II.11) with

$$\sigma^1 = dx, \qquad \sigma^2 = e^x \, dy, \qquad \sigma^3 = e^x \, dz. \qquad\qquad\text{(III.16)}$$

IV. Locally Boost Symmetric Spaces

These are case (B). The Jacobi identities for the triples (X_2, X_3, Y), (X_3, X_1, Y), and (X_1, X_2, Y) yield†

$$N_{12} = N_{13} = N_2 = N_3 = M_2 = M_3 = L_2 = L_3 = D = 0. \quad \text{(IV.1)}$$

The basis freedom and the remaining Jacobi identity again give (III.2). The conditions for a simply transitive subgroup are

$$L_1 - \alpha N_1 + 2\beta\gamma = 0,$$
$$\gamma(N_{23} - M_1 + \alpha) = 0, \qquad \beta(N_{23} + M_1 + \alpha) = 0. \qquad\text{(IV.2)}$$

† See the footnote on page 127.

The possibilities arising are

$$\beta = \gamma = 0, \qquad N_1\alpha = L_1, \qquad\qquad\qquad\text{(IV.3a)}$$

$$M_1 = 0, \quad N_{23} + \alpha = 0, \quad L_1 + N_{23}N_1 + 2\beta\gamma = 0, \qquad\text{(IV.3b)}$$

$$M_1 \neq 0, \quad N_{23} + M_1 + \alpha = 0, \quad \gamma = 0, \quad N_1\alpha = L_1, \qquad\text{(IV.3c)}$$

or

$$N_{23} - M_1 + \alpha = 0, \qquad \beta = 0, \qquad N_1\alpha = L_1. \qquad\text{(IV.3d)}$$

The simply transitive subgroup generators would obey

$$[\mathbf{Z}_2, \mathbf{Z}_3] = N_1\mathbf{Z}_1 - \gamma\mathbf{Z}_2 - \beta\mathbf{Z}_3,$$

$$[\mathbf{Z}_3, \mathbf{Z}_1] = (N_{23} - M_1 + \alpha)\mathbf{Z}_3, \qquad\qquad\text{(IV.4)}$$

$$[\mathbf{Z}_1, \mathbf{Z}_2] = (N_{23} + M_1 + \alpha)\mathbf{Z}_2.$$

There are the following cases.

B1. $M_1 = N_1 = 0$

Equation (II.14) shows there is a multiply transitive subgroup, which is of type VI_0 if $L_1 = 0$ and type VIII if $L_1 \neq 0$; the spaces on which it acts have curvature $-L_1$.

(a) $L_1 = 0$:   there are two one-parameter families of subgroups of type III, by (IV.3b), a one-parameter family of groups of type VI_0, by (IV.3a), and a subgroup of type I. The metric is

$$ds^2 = dt^2 + E^2\, dx^2 - 2F^2\, dy\, dz. \qquad\qquad\text{(IV.5)}$$

When $L_1 \neq 0$ there is always a one-parameter family of simply transitive subgroups of Bianchi type III, by (IV.3b). The metrics are

(b) $L_1 < 0$

$$ds^2 = dt^2 + E^2\, dx^2 - F^2(\sin y\, dz - dy)(\sin y\, dz + dy). \qquad\text{(IV.6)}$$

(c) $L_1 > 0$

$$ds^2 = dt^2 + E^2\, dx^2 - F^2(dy - \sin y\, dz)(dy + \sin y\, dz). \qquad\text{(IV.7)}$$

B2. $L_1 = M_1 = 0 \neq N_1$

There are no multiply transitive subgroups.

(a) $N_{23} = 0$:   there is a simply transitive subgroup of Bianchi type II, and, by (IV.3b), two one-parameter families of subgroups of type III. The metric is given by (II.12) and (III.10).

(b) $N_{23} \neq 0$:   there is a simply transitive subgroup of type VIII, and, by (IV.3b), a one-parameter family of subgroups of type III. The metric is (II.12) with

$$\sigma^1 = dx, \qquad \sigma^2 = du + \frac{\gamma}{\beta} e^{-\beta u}\, dv, \qquad \sigma^3 = e^{-\beta u}\, dv. \qquad \text{(IV.8)}$$

B3. $N_1 = L_1 = 0 \neq M_1$

There is a multiply transitive group of type VI_0 acting on flat timelike two surfaces, and there are a one-parameter family of simply transitive subgroups of type VI_h, including one of type V, by (IV.3a), and two one-parameter families of simply transitive subgroups of Bianchi type III, by (IV.3c) and (IV.3d). The metric is given by (II.12) and (III.16).

These metrics are related by complex transformations to their obvious counterparts in Section III.

V.   Locally Null-Rotation-Symmetric Spaces

These are case (C). The Jacobi identities for the triples $(\mathbf{X}_2, \mathbf{X}_3, \mathbf{Y})$, $(\mathbf{X}_3, \mathbf{X}_1, \mathbf{Y})$, and $(\mathbf{X}_1, \mathbf{X}_2, \mathbf{Y})$ give†

$$N_{12} = N_{23} = N_3 = M_1 = M_2 = L_2 = L_3 = D = 0, \quad N_{13} + N_2 = 0. \tag{V.1}$$

The remaining basis freedom can be used to set $N_1 = 0$, and the Jacobi identity for $(\mathbf{X}_1, \mathbf{X}_2, \mathbf{X}_3)$ gives

$$N_2 M_3 = 0. \tag{V.2}$$

The conditions for simply transitive subgroups are

$$L_1 + \alpha\gamma + N_2\gamma + M_3\beta - \beta^2 = 0,$$

$$\alpha M_3 + \beta(N_2 - \alpha) = 0, \qquad \alpha(N_2 - \alpha) = 0. \tag{V.3}$$

The possibilities arising are

$$N_2 = 0, \quad \alpha = 0, \quad \beta^2 - M_3\beta - L_1 = 0, \tag{V.4a}$$

$$N_2 \neq 0, \quad \alpha = \beta = 0, \quad L_1 + N_2\gamma = 0, \tag{V.4b}$$

or

$$\alpha = N_2, \quad M_3 = 0, \quad L_1 + 2N_2\gamma - \beta^2 = 0. \tag{V.4c}$$

† See the footnote on page 127.

The simply transitive subgroup would have generators obeying

$$[Z_2, Z_3] = -\gamma Z_1 + (\beta - M_3)Z_2 - N_2 Z_3,$$

$$[Z_3, Z_1] = M_3 Z_1 + (N_2 - \alpha)Z_2, \qquad \text{(V.5)}$$

$$[Z_1, Z_2] = (\alpha - N_2)Z_1.$$

We thus find that there is a multiply transitive subgroup, of Bianchi type II, in all cases except $M_3 = 0 \neq N_2$. All these cases with a multiply transitive subgroup can be put in the standard metric form admitting such a group, which has been shown by Barnes [20] to be the Petrov [11] form, the extra terms given by Defrise [9] being unneccessary; this metric is

$$ds^2 = \varepsilon^2(-2\,du\,dv + dy^2 - 2\phi\,dv^2) + dt^2, \qquad \text{(V.6)}$$

where ε and ϕ are functions of v and t. Integration of the Killing equations for the extra Killing vector yields

$$\varepsilon^2 = E^2(t)G^2(v), \qquad \phi = F(t)G^{-2}(v)(1 - M_3 v - L_1 v^2)^{-2}, \qquad \text{(V.7)}$$

where

$$G(v) = \exp \int \frac{(M_3 + L_1 v)dv}{1 - M_3 v - L_1 v^2}. \qquad \text{(V.8)}$$

For these metrics, $N_2 = 0$ and, by (V.4a), there is a simply transitive subgroup only if $M_1^2 + 4L_1 \geq 0$. The possible subcases are

C1. $M_3 = N_2 = 0$

(a) $L_1 > 0$:   there is a one-parameter family of simply transitive subgroups of type III.

(b) $L_1 = 0$:   there is a one-parameter family of groups of Bianchi type III and one group of Bianchi type I. This form includes plane waves.

(c) $L_1 < 0$:   no simply transitive subgroups G_3.

C2. $N_2 = 0 \neq M_3$

(a) $M_3^2 + 4L_1 > 0$:   there are two one-parameter families of simply transitive subgroups of Bianchi type VI_h, in general; if $L_1 = 0$, the types are IV and III, and if $L_1 = 2M_3^2$, VI_0 and III.

(b) $M_3^2 + 4L_1 = 0$:   there is a one-parameter family of subgroups of Bianchi type VI_h.

(c) $M_3^2 + 4L_1 < 0$:   no simply transitive subgroups G_3.

Although the cases with simply transitive G_3 can be put in the form
(II.13), the form (V.6) has the advantage of covering all the cases. It should be
noted that the obvious null tetrad in (V.6), which we shall use in Section VI,
is related to that of (II.13) by a v-dependent boost, and the Riemann tensor
will thus contain v-dependent terms that could be eliminated by a change of
null tetrad.

The case with no multiply transitive subgroup is

C3. $M_3 = 0 \neq N_2$

Whatever the value of L_1, (V.4b) gives a simply transitive group of Bianchi
type VIII, and (V.4c) gives a one-parameter family of groups of type III.
[L_1 essentially gives an initial value for F in (II.13).] The metric is

$$ds^2 = dt^2 + E^2[dy^2 - 2e^y \, dv(du + Fe^y \, dv)]. \tag{V.9}$$

VI. The Ricci and Weyl Tensors

Cases (A1) and (A2)

From (III.15), using the tetrad $\{E[dx + lf(y)\,dz], F\,dy, Ff'(y)\,dz, dt\}$, we
find the only nonzero Ricci tensor components are

$$R_{11} = \frac{E^{\cdot\cdot}}{E} + 2\left(\frac{E^{\cdot}F^{\cdot}}{EF} + \frac{lE^2}{4F^4}\right),$$

$$R_{22} = R_{33} = \frac{k}{F^2} + e\left(\frac{F^{\cdot\cdot}}{F} + \frac{E^{\cdot}F^{\cdot}}{EF} + \frac{F^{\cdot 2}}{F^2} - \frac{lE^2}{2F^4}\right), \tag{VI.1}$$

$$R_{44} = -\frac{E^{\cdot\cdot}}{E} - \frac{2F^{\cdot\cdot}}{F}.$$

The Ricci tensor is therefore of type [(11)1,1] (or its specializations). The
type D Weyl tensor has value characterized by

$$\psi_2 = \frac{1}{6}\,e\left(\frac{F^{\cdot\cdot}}{F} - \frac{E^{\cdot\cdot}}{E} + \frac{E^{\cdot}F^{\cdot}}{EF} - \frac{F^{\cdot 2}}{F^2} + \frac{lE^2}{F^4}\right) - \frac{k}{F^2} - \frac{ilE^2}{2F^2}\left(\frac{E^{\cdot}}{E} - \frac{F^{\cdot}}{F}\right), \tag{VI.2}$$

(cp. [14]); it becomes conformally flat if $\psi_2 = 0$, for example, if $l \neq 0$, when
$e = 1$ and $E = F$.

Case (A3)

Using the complex null tetrad $(m, \bar{m}, l, k)$ given by $[\sqrt{\tfrac{1}{2}}F(\sigma^2 + i\sigma^3),$ $\sqrt{\tfrac{1}{2}}F(\sigma^2 - i\sigma^3), \sqrt{\tfrac{1}{2}}e(dt - E\sigma^1), \sqrt{\tfrac{1}{2}}(dt + E\sigma^1)]$, the nonzero curvature components are given by

$$\phi_{00} = \frac{2\dot{F}}{EF} - E^{-2} - \frac{(1-e)\dot{F}^2}{F^2} - e\left(\frac{\ddot{F}}{F} - \frac{\dot{E}\dot{F}}{EF} + \frac{2\dot{E}^2}{E^2}\right),$$

$$\phi_{11} = -\frac{1}{2}\left(\frac{\dot{E}}{E}\right)^{\cdot} + \frac{1}{2}e\left(\frac{\dot{F}^2}{F^2} - \frac{\dot{E}^2}{E^2} - E^{-2}\right),$$

$$\phi_{22} = -\frac{1}{2}\left(\frac{\dot{F}}{F} + E^{-1}\right)^2 + \frac{\dot{E}}{E}\left(\frac{\dot{F}}{F} + E^{-1}\right) + \frac{1}{2}e\left(\frac{\dot{F}}{F} + E^{-1}\right)^{\cdot}, \quad \text{(VI.3)}$$

$$6\Lambda = \left(\frac{\dot{F}}{F}\right)^{\cdot} + \frac{1}{2}\left(\frac{\dot{E}}{E}\right)^{\cdot} + e\left(\frac{3\dot{F}^2}{2F^2} - \frac{3}{2E^2} + \frac{\dot{E}\dot{F}}{EF} + \frac{\dot{E}^2}{2E^2}\right),$$

$$6\psi_2 = \left(\frac{\dot{F}}{F}\right)^{\cdot} - \left(\frac{\dot{E}}{E}\right)^{\cdot} + e\left(\frac{\dot{E}^2}{E^2} - \frac{\dot{E}\dot{F}}{EF}\right).$$

The Weyl tensor is of type D unless $\psi_2 = 0$ (when it is conformally flat), and the Ricci tensor is of type $[(11)1,1]$ (or its specializations) unless exactly one of ϕ_{00} and ϕ_{22} is zero, in which case it is $[(11),2]$ if $\phi_{11} \neq 0$ or $[(11,2)]$ if $\phi_{11} = 0$.

Case (B)

There is always a simply transitive group (IV.4), and, using the related reciprocal group generators in (II.12), and taking the obvious null tetrad $[\sqrt{\tfrac{1}{2}}(E\sigma^1 + i\,dt), \sqrt{\tfrac{1}{2}}(E\sigma^1 - i\,dt), F\sigma^2, F\sigma^3]$, one finds the nonzero curvature components to be

$$\phi_{20} = \bar{\phi}_{02} = \frac{1}{2}\left[\left(\frac{\ddot{F}}{F} - \frac{\dot{E}\dot{F}}{EF}\right) - \frac{1}{2}\left(\frac{N_1^2 E^2}{4F^4} + \frac{M_1^2}{E^2}\right)\right] - 2i\frac{M_1\dot{F}}{EF},$$

$$4\phi_{11} = -\frac{\ddot{E}}{E} + \frac{\dot{F}^2}{F} + \frac{M_1^2}{E^2} - 3\left(\frac{N_1 E}{2F^2}\right)^2 + \frac{(N_1 N_{23} + L_1)}{F^2},$$

$$6\Lambda = -\frac{\ddot{F}}{F} - \frac{1}{2}\frac{\ddot{E}}{E} - \frac{\dot{E}\dot{F}}{EF} + \frac{1}{2}\left(\frac{N_1 E}{2F^2}\right)^2 - \frac{3M_1^2}{2E^2} - \frac{(N_1 N_{23} + L_1)}{2F^2},$$

$$6\psi_2 = \left[\left(\frac{\dot{E}}{E} - \frac{\dot{F}}{F}\right)^{\cdot} + \frac{\dot{E}}{E}\left(\frac{\dot{E}}{E} - \frac{\dot{F}}{F}\right) + \frac{(N_1 N_{23} + L_1)}{F^2}\right]$$
$$+ \frac{1}{2}i\frac{N_1 E}{F^2}\left(\frac{\dot{E}}{E} - \frac{\dot{F}}{F}\right).$$

In general, the Petrov type is D, and the Ricci tensor type is $[11(1,1)]$, but specializations are possible.

Cases (C1) and (C2)

Taking the metric (V.6), and the null tetrad $(m, \bar{m}, l, k)$ given by $[\sqrt{\tfrac{1}{2}}(dt + i\varepsilon\, dy), \sqrt{\tfrac{1}{2}}(dt - i\varepsilon\, dy), \varepsilon(du + \phi\, dv), \varepsilon\, dv]$, the nonzero curvature components [remembering (V.7)] are

$$2\phi_{11} = \phi_{02} = \bar{\phi}_{20} = -\frac{1}{2}\left(\frac{\dot{\varepsilon}}{\varepsilon}\right)^{\!\cdot},$$

$$2\phi_{22} = -\left(\frac{\varepsilon''}{\varepsilon^3} - \frac{2\varepsilon'^{\,2}}{\varepsilon^4} + \ddot{\phi} + \frac{3\dot{\varepsilon}\,\dot{\phi}}{\varepsilon}\right),$$

$$2\psi_4 = \left(\frac{\varepsilon''}{\varepsilon^3} - \frac{2\varepsilon'^{\,2}}{\varepsilon^4} - \ddot{\phi} - \frac{\dot{\varepsilon}\,\dot{\phi}}{\varepsilon}\right),$$

$$\Lambda = -\frac{1}{4}\left(\frac{\ddot{\varepsilon}}{\varepsilon} + \frac{\dot{\varepsilon}^{\,2}}{\varepsilon^2}\right),$$

(VI.5)

where the prime denotes $\partial/\partial v$ and, as usual, the dot $\partial/\partial t$. (VI.5) agrees with Barnes' result [20].

Case (C3)

Using the null tetrad $[\sqrt{\tfrac{1}{2}}(E\, dy + i\, dt), \sqrt{\tfrac{1}{2}}(E\, dy - i\, dt), E(du + Fe^y\, dv), Ee^y\, dv]$ the nonzero curvature components are given by

$$\phi_{11} = \tfrac{1}{2}\phi_{02} = \Lambda + \left(\frac{1}{8E^2} + \frac{\dot{E}^{\,2}}{2E^2}\right),$$

$$2\phi_{02} = \left(\frac{\dot{E}}{E}\right)^{\!\cdot} - \frac{1}{4E^2},$$

$$2\phi_{22} = \ddot{F} + \frac{3\dot{E}\dot{F}}{E} + \frac{2F}{E^2},$$

(VI.6)

$$\psi_4 = \tfrac{1}{2}\ddot{F} + \frac{F}{E^2} + \frac{\dot{E}\dot{F}}{2E} + \frac{i}{2E}\left(3\dot{F} + \frac{\dot{E}F}{E}\right).$$

For both case (C) metrics, the Weyl tensor is of type N, and the Ricci tensor of type $[1(1,2)]$ or specializations, except when $\phi_{22} = 0$, in which case the Ricci tensor may be type $[1(11,1)]$ or specializations.

VII. Further Remarks

Table I shows the comparison between the present list of metrics and that given by Petrov [11]. Petrov's list is indexed by the Kruchkovich–Petrov type of the group G_4 (KP type). My reasons for not following Petrov are (i) the KP classification into eight types is not fine enough and needs complicated subclassification,† (ii) the KP classification contains some overlapping classes,† (iii) the numbering leads to confusion with the Bianchi types of G_3, to which it is not related; (iv) the same group may have different actions, and (v) it throws no light on the geometrical nature of the isotropy. The table gives the metrics by reference to the English edition of Petrov's work [11]; some discrepancies with the Russian and German editions were noted by Ray and Zimmerman [21]. The only discrepancy within the English edition's list is that the metric (32.12) is stated to have KP type VI_4.

Cahen and Defrise [8] gave all Petrov type D metrics with isotropy. (A1) and (A3) are contained in their metric (2.3b), and (A2) in (2.6) and (2.8); (B1) and (B3) are included in (2.9), and (B4) in (2.11). Defrise [9] considered the isotropic type N metrics and gave (V.6) in different coordinates, and (C3) with $\phi_{02} = 0$. The locally rotationally symmetric metrics [case (A)] were all given by Ellis and Stewart [6,7].

Ray and Zimmerman [21] considered dust and tachyonic dust solutions for Petrov's metrics (32.3)–(32.12). Their conclusions concerning the

TABLE I[a]

Metric	A1a	A1b	A1c	A2a	A2b	A2c	A3
KP G_4 type	VI_4	VII	VIII	III ($q = 0$)	VIII	VII	V
Petrov's metric	(32.11) $\varepsilon = 1$	(32.7) $e_3 = 1$	(32.9)	(32.4)	(32.10)	omitted	(32.6)

Metric	B1a	B1b/c	B2a	B2b	B3
KP G_4 type	VI_1	VII	I ($c = 0$)	VII	IV
Petrov's metric	(32.11) $\varepsilon = -1$ $e_4 = 1$	(32.7) $e_3 = 1$	(32.3)	omitted	(32.5)

Metric	C1a & C2a	C1b	C1c & C2c	C2b	C2b
KP G_4 type	I	VI_2 ($K = 0, \varepsilon = 1$) and VI_3	III	II	VII
Petrov's metric	(32.14)	(32.12)	(32.16)	(32.15)	(32.8)

[a] For explanation, see text.

† I hope to publish a note on these matters elsewhere.

admissibility of spacelike or timelike four-velocities follow much more easily from the present approach, since the four-velocity must be invariant under the isotropy and hence may be either spacelike or timelike in case (A), but only spacelike in cases (B) and (C). Incidentally, their remarks about case (A2a) seem to be in error: There are known dust and vacuum solutions, and the dust could be tachyonic (see [1,24]).

Abdel-Megied [22] has classified the Riemann tensor for Petrov's metrics (32.3)–(32.9), except (32.7), and considered degeneracies.

The case (A) spatially homogeneous metrics have been widely considered as cosmological models (cf. [3–5]). The present method gives an easy proof of Kantowski's result that case (A1c) gives the only spatially homogeneous cosmologies with no simply transitive G_3 (cf. [18,23]). The conformally flat cases with $E = F$ in (A1a), (A2b), and (A3) with $e = 1$, are the well-known Robertson–Walker metrics.

A full account of all occurrences in the literature of the metrics given here, and of the related exact solutions, would take far too much space to give here; the references must, perforce, suffice. I would, however, like to point out that the exact vacuum solutions of Einstein's equations for (A1a) and (A1b) were given by Taub [1] as was that for (A2); the latter is the famous Taub–NUT metric, in which portions with $e = 1$ and $e = -1$ are joined across null surfaces; the same interesting global property can occur in other case (A) solutions.

Acknowledgments

I am grateful to Dr. D. Kramer and Dr. H. Stephani of the University of Jena, and to Barnes [20], Ray and Zimmerman [21], Abdel-Megied [22], and Collins [23] for stimuli that combined to suggest this work, and to Dr. B. G. Schmidt for helpful conversations.

References

[1] Taub, A. H., *Ann. Math.* **53**, 472 (1951).
[2] Eisenhart, L. P., "Continuous Groups of Transformations." Princeton Univ. Press, Princeton, New Jersey, 1933. (Reprinted, Dover, New York, 1961.)
[3] MacCallum, M. A. H., *in* "Cargese Lectures" Vol. 6, (E. Schatzman, ed.). p. 61–174. Gordon & Breach, New York, 1973.
[4] Ryan, M. P., Jr., and Shepley, L. C., "Homogeneous Relativistic Cosmologies." Princeton Univ. Press, Princeton, New Jersey, 1975.
[5] MacCallum, M. A. H., *in* "Physics of the Expanding Universe" (M. Demianski, ed.), Lecture Notes in Physics, Vol. 109, pp. 1–59. Springer-Verlag, Berlin, 1979.
[6] Ellis, G. F. R., *J. Math. Phys.* **8**, 1171 (1967).
[7] Stewart, J. M., and Ellis, G. F. R., *J. Math. Phys.* **9**, 1072 (1968).
[8] Cahen, M., and Defrise, L., *Commun. Math. Phys.* **11**, 56 (1968).

[9]   Defrise, L., Ph.D. Thesis, Univ. Libre de Bruxelles, Brussels, 1969.
[10] Siklos, S. T. C., "Algebraically Special Homogeneous Space-times" Prepr. Oxford Univ. 1978.
[11] Petrov, A. Z., "Einstein Spaces." Pergamon, Oxford, 1969.
[12] Schmidt, B. G., Ph.D. Thesis, Univ. of Hamburg, Hamburg, 1968.
[13] Schmidt, B. G., *Gen. Relativ. Grav.* **2**, 105 (1971).
[14] Newman, E. T., and Penrose, R., *J. Math. Phys.* **3**, 566 (1962).
[15] Ehlers, J., and Kundt, W., *in* "Gravitation: An Introduction to Current Research" (L. Witten, ed.), pp. 48-101. Wiley, New York, 1962.
[16] Hall, G. S., *J. Phys. A* **9**, 541 (1976).
[17] Schmidt, B. G., *Z. Naturforsch. Teil A* **22a**, 1351 (1967).
[18] Kantowski, R., and Sachs, R. K., *J. Math. Phys.* **7**, 443 (1966).
[19] Newman, E. T., Tamburino, L. A., and Unti, T., *J. Math. Phys.* **4**, 915 (1963).
[20] Barnes, A., *J. Phys. A.* **12**, 1493 (1979).
[21] Ray, J. R., and Zimmerman, J. C., *J. Math. Phys.* **18**, 881 (1977).
[22] Abdel-Megied, M., "Classification of Ricci Tensor in Spatially-Homogeneous Space-times," Prepr. Univ. of Riyad, Riyad, 1978.
[23] Collins, C. B., *J. Math. Phys.* **18**, 2116 (1977).
[24] Collins, C. B., *Commun. Math. Phys.* **23**, 137 (1971).

10

The Gravitational Waves That Bathe the Earth: Upper Limits Based on Theorists' Cherished Beliefs[†‡]

Mark Zimmermann[§] and Kip S. Thorne

W. K. Kellogg Radiation Laboratory
California Institute of Technology
Pasadena, California

Abstract

On the basis of our cherished beliefs about the structure of the Universe and the theory of gravitation, we derive theoretical upper limits on the strengths of the gravitational waves that bathe the Earth. Separate limits are presented, as functions of frequency, for waves from extragalactic sources and for waves from inside our own galaxy; and in each case, for discrete sources (bursters, transient sources, and monochromatic sources) and for a stochastic background due to unresolved sources. An observation of gravitational waves exceeding these limits would be disturbing (and exciting), since it would require a modification of one or more generally accepted assumptions about the astrophysical universe or the nature of gravity.

I. Introduction

During the past two decades general relativity theory has had an increasingly strong impact on astrophysics—first in the theory of quasars;

† This paper is dedicated to our good friend and colleague, Abraham H. Taub, on the occasion of his retirement from the University of California at Berkeley, Berkeley, California.

‡ Supported in part by the National Aeronautics and Space Administration [NGR 05-002-256] and by the National Science Foundation [AST76-80801 A02].

§ Present address: IDA/SED, 400 Army Navy Drive, Arlington, Virginia 22202.

then in cosmology, pulsars, compact x-ray sources, and the search for black holes. We hope for an even stronger impact in the future, when gravitational waves open up a new "window" onto the Universe—a window in which general relativity will play an absolutely essential role.

The efforts of experimenters to develop gravitational-wave detectors of ever-increasing sensitivity have been described in a number of recent review articles [1–4]. As these efforts proceed, it is useful to have theoretical "benchmarks" against which to gauge their progress. Such benchmarks are of three major types. The first type, as sensitivities improve, are "nihil obstat" upper limits on the strengths of the waves. An observation of waves above these limits would overturn one or more cherished beliefs about either the structure of the Universe or the physical laws governing gravitational radiation. Type-two benchmarks are at a level where the best estimates of plausible astrophysical sources indicate that something should be seen. Observations at these sensitivities are sure to give significant astronomical information; even if no waves are detected, many otherwise acceptable models will be eliminated. Type-three benchmarks are the absolute minimum gravitational-wave strengths consistent with other astronomical observations. A failure to see waves below these limits would be as serious a matter as observations of waves above the type-one limits; in either case, something is radically wrong with the theory of gravitation or with conventional astrophysical wisdom.

Type-two and type-three benchmarks have been reviewed in several recent articles [5,6]. The purpose of this article is to set forth benchmarks of the first type—"cherished-belief" upper limits on gravitational-wave strengths.

In Section II we list and discuss the cherished beliefs on which our limits are based. In Section III we derive, from those cherished beliefs, upper limits on the strength of any stochastic background of gravitational waves that might bathe the Earth—both a limit on waves from unresolved sources in our galaxy, and a limit on extragalactic waves. We also describe scenarios that could lead to these upper limits. In Section IV we derive similar upper limits on waves from discrete sources including bursters, transient sources, and monochromatic sources. Again there are separate upper limits for sources in our own galaxy and extragalactic sources. For the case of broadband bursts, we also describe a scenario that could lead to the galactic upper limits.

Throughout we shall restrict attention to gravitational-wave frequencies in the domain of current experimental interest: 10^{-4} Hz $\lesssim f \lesssim 10^{+4}$ Hz. The lower limit 10^{-4} Hz is dictated by the technology of gravitational-wave detectors [1–4]—in particular, the round-trip radio-wave travel time to spacecraft at reasonable distances (e.g., Jupiter). The upper limit 10^{+4} Hz is dictated by our *cherished belief* [5] that the only highly efficient sources of gravitational waves in the Universe today are objects near their Schwarzschild radii—neutron stars and black holes of stellar mass and larger—and that these objects cannot radiate significantly at frequencies above $f_{max} \simeq 10^{+4}$ Hz.

The notation used in our discussion is summarized in the Appendix. A more detailed discussion of each parameter is given at the point in the text where it is first introduced.

II. Cherished Beliefs

The cherished beliefs, on which we base our limits, are of two types: beliefs about the astrophysical structure of the Universe (Section II.A), and beliefs about the physical laws governing gravitational radiation (Section II.B).

A. *The Structure of the Universe*

Our first cherished belief is the *cosmological principle* that *we do not live in a special time or place in the Universe*—except for being inside a local density enhancement, the galaxy. The cosmological principle implies that, on the average, sources of gravitational waves are no more luminous now than they have been (and will be) for a Hubble time $T_H = 1 \times 10^{10}$ years, the only time scale available. It also implies that the nearest source is at a typical distance from us, neither fortuitously near nor far. (For objects of number density n in Euclidean three-space, the mean distance to the nearest one is $0.55396 \ldots n^{-1/3}$; over 90% of the time, the nearest is between $0.2n^{-1/3}$ and $0.9n^{-1/3}$. We shall use $0.5n^{-1/3}$ as the distance to the nearest source throughout this paper.)

Our next cherished belief is that there is *no significant amount of "relict," primordial gravitational radiation* bathing the Earth—more precisely, that all the significant sources of gravitational waves are at cosmological red-shifts $z \lesssim 3$. This is as much a simplifying assumption as a cherished belief: Although semiplausible models of the early Universe give only modest amounts of gravitational radiation [7,8] (amounts well below the upper limits of this paper), we are so ignorant about the early Universe that it is hard to place *firm* upper limits on the waves from there, except the obvious limit that their total energy density not exceed by much the density required to close the Universe. The closure limit will follow from our other cherished beliefs without explicitly assuming it.

The cosmological principle, plus the belief in "no primordial waves," allows us to approximate the Universe by a very simple model that is accurate to within an order of magnitude in energies (a factor of three in gravitational-wave amplitudes). In this model the expansion of the Universe is ignored, space is regarded as Euclidean, the Universe is regarded as extending outward from Earth in all directions to a Hubble distance $R_H = cT_H = 1 \times 10^{10}$ light years, within this distance the smeared-out mass density of potential gravitational-wave sources is regarded as constant and as equal to the "closure density" $\rho_u = (c^2/G)(3/8\pi)R_H^{-2} = 2 \times 10^{-29} \text{ g/cm}^3$, and outside R_H the density drops to zero (cosmological cutoff on sources). Our use of the closure density for ρ_u does not mean that we believe in this value, but rather that this is a reasonable upper limit and will thus give rise to the largest possible limits on gravitational-wave strengths. The galaxy we shall model as a region of constant, enhanced mean mass density, $\rho_g = 2 \times 10^{-24} \text{ g/cm}^3$ (no radial structure), and of spherical shape with radius $R_g = 60,000$ light years and with the Earth located (roughly) at its center. The numbers for our galaxy take account of a now popular galactic halo with total mass $M_g = (4\pi/3)R_g^3\rho_g \simeq 1 \times 10^{12} M_\odot$ and radius $R_g \simeq 60,000$ light years [9–11].

Our third cherished belief is that *within our galaxy no single, coherently radiating object has mass in excess of* $M_{max} \simeq 10^8 M_\odot$ [12]. This is a very generous upper limit. We make no assumption about the maximum mass of extragalactic objects.

Our fourth cherished belief is that the dominant sources of gravitational waves have *no significant beaming* of their radiation. In principle, strong beaming can occur—e.g., in waves from ultrarelativistic collisions of astrophysical objects [13,14], in waves from sources with gravitational lens properties [15,16], and in waves from carefully contrived directional antennas [17]. However, we do not know of any type of hypothetical strong-beaming source that is likely to make up a significant fraction of the mass density of the galaxy or Universe. Moreover, our limits are fairly insensitive to the no-beaming assumption: A simple geometrical analysis in flat space shows that, if sources beam their energy into a solid angle $\Omega < 4\pi$, and if the Earth is located randomly relative to the beams, then the expected energy flux from the nearest visible object increases only as $(\Omega/4\pi)^{-1/3}$, and the expected total flux from all sources out to some fixed cutoff radius remains constant. (On the other hand, as Lawrence [15], Misner [18], Jackson [16], and others have argued, there *could* be an object at the center of our galaxy that preferentially beams its radiation into the galactic plane where we lie. Our no-beaming assumption rules this out.)

Our fifth cherished belief is that narrow band sources of gravitational waves($\Delta f \ll f$) have their *frequencies f distributed randomly over a bandwidth* $\Delta f \gtrsim f$.

B. *The Physical Laws Governing Gravitational Radiation*

We take our cherished beliefs about gravitational-wave theory from general relativity, though most other relativistic theories of gravity would lead to similar beliefs. Our beliefs are expressed in order-of-magnitude form.

Consider a source of mass M, which radiates gravitational waves coherently. (Examples: A pulsating star, a binary star system, two colliding black holes.) If small parts of the source produce waves that superpose incoherently, those parts must be regarded as separate sources. (Example: for the thermal bremsstrahlung radiation produced by collisions of electrons and ions inside the sun, the source is not the entire sun but rather a single colliding electron–ion pair.) Let f be a frequency at which the source radiates significantly. Our first cherished belief is an *upper limit on the frequency f, for a given source mass M* [5]:

$$f \lesssim \frac{c^3/G}{2\pi M} \simeq \frac{30,000 \text{ Hz}}{M/M_\odot}. \tag{1}$$

This limit corresponds to a belief that the characteristic time scale $(2\pi f)^{-1}$ of the coherent waves must exceed the light-travel time across half the Schwarzschild radius of the source GM/c^3. This limit can be violated in sources with significant beaming—e.g., sources with ultrarelativistic internal velocities [13,14]; but we have ruled out such sources. We strongly doubt that coherent, nonbeaming sources can violate this limit. For example, typical events involving black holes (births, collisions, infall of matter) produce waves of frequency $f \sim 10,000 \text{ Hz } (M/M_\odot)^{-1}$ [5,19–21], with a very rapid falloff of intensity above $f = 30,000 \text{ Hz } (M/M_\odot)^{-1}$.

In general relativity and other similar theories, a source with negligible beaming gives rise predominantly to quadrupole radiation. The luminosity of such a source is given by Einstein's [22] quadrupole formula

$$L = \frac{1}{5}\left(\frac{G}{c^5}\right)\left(\sum_{j,k} \frac{\partial^3 I_{jk}}{\partial t^3}\right)^2,$$

where the third time derivative of the quadrupole moment, expressed in terms of the coherent source's mass M, radius R, and frequency f, is

$$\frac{\partial^3 I_{jk}}{\partial t^3} \lesssim MR^2(2\pi f)^3 \lesssim 2\pi Mfc^2.$$

Here we have used the relation, for a coherent source,

$$(2\pi f R) \simeq (\text{internal velocity of source}) \lesssim c.$$

Combining these relations we obtain a cherished belief about the *maximum luminosity that a source of mass M and frequency f can produce*

$$L \lesssim \frac{G}{c}(2\pi M f)^2 \simeq (4 \times 10^{50} \text{ erg/sec})\left(\frac{M}{M_\odot}\right)^2 \left(\frac{f}{1 \text{ Hz}}\right)^2. \qquad (2)$$

Note that when $f = f_{\max} = (c^3/G)(2\pi M)^{-1}$, then $L \lesssim L_{\max} \simeq c^5/G \approx (4 \times 10^{59} \text{ erg/sec})$—a limit which, so far as we know, was first suggested by Dyson [23].

In our analysis we shall idealize our typical source as radiating gravitational waves in a series of outbursts separated by quiescent periods. Let N denote the total number of outbursts, τ_* the mean duration of each outburst, and L the average luminosity during each outburst. Our next cherished belief is that, in the source's entire lifetime, *the total energy radiated cannot exceed the total mass-energy Mc^2 of the source*

$$NL\tau_* \leq Mc^2. \qquad (3)$$

In describing the gravitational waves arriving at Earth we shall use, at various times, four different measures of wave strength: *First*, in describing waves from discrete sources we shall use a mean value h for the dimensionless gravitational-wave amplitude at the frequency f in a bandwidth $\Delta f \simeq f$:

$$h \simeq \langle [h_+(t)]^2 + [h_\times(t)]^2 \rangle^{1/2}.$$

Here the average $\langle\ \rangle$ is over the time τ_* that the source is on; and $h_+(t)$ and $h_\times(t)$ are the dimensionless amplitudes for the two orthogonal modes of polarization, which for a source in the z direction determine the transverse-traceless part of the metric perturbation via

$$h^{TT} = h_+(t-z)[\mathbf{e}_x \otimes \mathbf{e}_x - \mathbf{e}_y \otimes \mathbf{e}_y] + h_\times(t-z)[\mathbf{e}_x \otimes \mathbf{e}_y + \mathbf{e}_y \otimes \mathbf{e}_x].$$

We presume that h_+ and $h_\times$ have been sent through a bandpass filter of frequency f and bandwidth $\Delta f \simeq f$. For monochromatic waves, $h_+(t) = A_+ \cos(2\pi f t + \phi_+)$ and $h_\times(t) = A_\times \cos(2\pi f t + \phi_\times)$, our definition of h gives $h = [\frac{1}{2}(A_+^2 + A_\times^2)]^{1/2}$. *Second*, for discrete sources we shall also use the total flux of energy $\mathscr{F}$ at the frequency f and in the bandwidth $\Delta f \simeq f$. We shall assume (cherished belief!) the general relativistic *relationship between $\mathscr{F}$ and h*:

$$\mathscr{F} = \frac{c^3}{G}\frac{\pi}{4}f^2 h^2 = \left(0.03 \frac{\text{erg}}{\text{cm}^2 \text{ sec}}\right)\left(\frac{f}{1 \text{ Hz}}\right)^2 \left(\frac{h}{10^{-20}}\right)^2. \qquad (4)$$

Third, in describing stochastic background radiation, we shall use the energy flux per unit frequency $\mathscr{F}_f$ (flux density; erg/cm^2 sec Hz), which our cherished beliefs imply will be independent of the orientation of our unit surface area. *Fourth*, for the stochastic background we shall also use an amplitude $\tilde{h}(f)$ (dimensions $Hz^{-1/2}$), which is defined in analogy with Eq. (4) by

$$\mathscr{F}_f = \frac{c^3}{G}\frac{\pi}{4}f^2\tilde{h}^2 = \left(0.03 \frac{erg}{cm^2 \ sec \ Hz}\right)\left(\frac{f}{1 \ Hz}\right)^2\left(\frac{\tilde{h}}{10^{-20} \ Hz^{-1/2}}\right)^2. \quad (5)$$

The square of $\tilde{h}$, roughly speaking, is the spectral density of the gravitational-wave amplitude $h(t)$. The stochastic background will produce in a broadband gravitational-wave detector a spectral density of strain $(\Delta l/l)^2_f = \alpha\tilde{h}^2$, where α is a factor of order unity that depends on the detailed construction of the detector.

In relating the strengths of the waves at earth to the luminosity L of a source at distance r, we shall assume *energy conservation* (cherished belief!)

$$\mathscr{F}_{\text{due to one source}} = \frac{L}{4\pi r^2}, \quad (6)$$

and we shall assume that *gravitational waves propagate at the speed of light* (cherished belief!).

III. Upper Limits on Stochastic Background

From the cherished beliefs of Section II, one can derive the upper limits on a stochastic background of gravitational radiation shown in Fig. 1. In Section III.A we explain the origin of the limit for extragalactic radiation; in Section III.B we explain the galactic limit.

A. *Extragalactic Radiation*

Consider a specific frequency f at which the background is strong, and let Δf be the bandwidth about f over which the specific flux $\mathscr{F}_f$ is roughly constant. For a background due to broadband sources, by definition of "broadband," we have $\Delta f \gtrsim f$. For a background due to superposed narrowband sources, the last cherished belief of Section II.A ("frequencies distributed randomly over a bandwidth $\gtrsim f$") implies $\Delta f \gtrsim f$. Thus, in either case the background is roughly constant over $\Delta f \gtrsim f$, but it can drop off fairly rapidly at both ends of this band.

An upper limit on $\mathscr{F}_f$, for extragalactic background, follows from our cherished beliefs that (i) the total energy radiated by all sources cannot exceed the sum of the masses of those sources; and (ii) we do not live at a special place or time, so that the total gravitational-wave energy must be spread roughly uniformly over the entire Universe and the energy density at

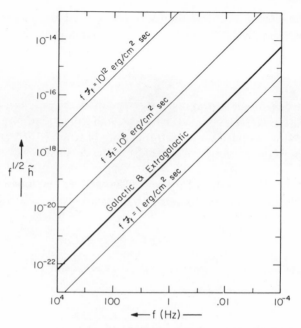

Fig. 1 Upper limits on a stochastic background of gravitational radiation at Earth [Eqs. (7) and (8)]. The limit for radiation from sources in our galaxy is approximately the same as the limit for extragalactic radiation. For notation, see Eq. (5) of text and associated discussion.

Earth must be roughly the same as the average energy density in the Universe. These beliefs imply a total energy density in background radiation at Earth less than or of order the total mass-energy density of the Universe:

$$\left(\begin{array}{c}\text{background}\\\text{energy density}\end{array}\right) = \frac{4\mathscr{F}_f\,\Delta f}{c} \lesssim \rho_u c^2.$$

(The factor 4 comes from integrating over all directions.) Combining this with the bandwidth requirement $\Delta f \gtrsim f$, we obtain the limit

$$\mathscr{F}_f \lesssim \frac{1}{4}\left(\frac{\rho_u c^3}{f}\right) \simeq \left(100\,\frac{\text{erg}}{\text{cm}^2\,\text{sec Hz}}\right)\left(\frac{f}{1\,\text{Hz}}\right)^{-1}, \tag{7a}$$

which corresponds to a wave amplitude

$$f^{1/2}\tilde{h} \lesssim 6 \times 10^{-19}(f/1\,\text{Hz})^{-1} \tag{7b}$$

[cf. Eq. (5)]. These limits are shown in Fig. 1.

These extragalactic limits are widely accepted and often discussed in the astrophysical literature (see, e.g. [24]).

The upper limit (7) can be achieved, within the framework of our cherished beliefs, in a variety of ways. For example, at any frequency $f \lesssim 10,000$ Hz the following scenario is allowed, though not likely: Early in the evolution of the Universe a sizable fraction of the Universe's mass might have gone into black-hole binary systems of mass $M \sim (c^3/G)(2\pi f)^{-1}$. Under the action of gravitational radiation reaction, the holes in each binary will spiral together, releasing a sizable fraction of their mass M in a final burst of broadband radiation of frequency f and duration $\tau_* \sim (2\pi f)^{-1}$. These bursts must be randomly distributed over the volume of the Universe and over the Hubble time, so that the average number of bursts occurring at any given time is

$$\mathcal{N} = \frac{(4\pi/3)\rho_u R_H^3}{M} \frac{\tau_*}{T_H} \sim 1$$

(cf. relations in Section A of the Appendix). This is also the average number of bursts passing Earth at each moment of time; and these bursts give rise to background radiation near the upper limit (7).

One can also achieve these upper limits by a superposition of many bursts with lower individual intensities and longer individual durations.

B.   Galactic Radiation

For galactic background radiation, as for extragalactic, the bandwidth over which $\mathcal{F}_f$ is large must be $\Delta f \gtrsim f$. The radiation must be spread roughly uniformly over the interior of the galaxy ("no special place"), so that the total radiation energy in the galaxy is $(4\mathcal{F}_f \, \Delta f)(4\pi/3)(R_g^3/c)$. This radiation energy will escape from the galaxy in a time R_g/c and must be replenished by source emission in that time. The total energy emitted during the Hubble time T_H is thus $(cT_H/R_g) \times$ (total energy now in galaxy); and this cannot exceed the total mass-energy of the galaxy $(4\pi/3)R_g^3\rho_g c^2$. Combining these constraints we obtain the upper limit

$$\mathcal{F}_f \lesssim \frac{1}{4} \frac{R_g \rho_g c^2}{f T_H} \simeq 100 \frac{\text{erg}}{\text{cm}^2 \sec \text{Hz}} \left(\frac{f}{1 \text{ Hz}}\right)^{-1}, \tag{8a}$$

which corresponds to a wave amplitude

$$f^{1/2}\tilde{h} \lesssim 6 \times 10^{-19}(f/1 \text{ Hz})^{-1}. \tag{8b}$$

Note that this is the same order-of-magnitude limit as we obtained for extragalactic radiation! It is the same by virtue of the coincidence (or is it a coincidence?) that the closure density ρ_u and Hubble distance $R_H = cT_H$ of the

Universe, and the density ρ_g and radius R_g of the galaxy satisfy

$$\rho_u \sim \rho_g \left(\frac{R_g}{R_H}\right).$$

The upper limit (8) can be achieved, within the framework of our cherished beliefs, by putting the bulk of the mass of our galaxy into objects of mass $\sim M$ [with $M \lesssim (c^3/G)(2\pi f)^{-1}$ and $M \lesssim M_{max} = 10^8 \, M_\odot$], which radiate away all their mass in bursts of mean frequency f, duration τ_*, and luminosity $L = Mc^2/\tau_*$ [with $L \lesssim (G/c)(2\pi Mf)^2$ and $\tau_* \lesssim T_H$]. The locations of these objects and the epoch of their emission must be randomly distributed through the galaxy; so the number of "on" sources contributing to $\mathscr{F}_f$ at Earth at any given time will be $\mathscr{N} \simeq (M_g/M)(\tau_*/T_H)$ [where $\mathscr{N} \gtrsim 1$ so that experimenters will see a background rather than individual events]. The mass M and burst duration τ_* can be chosen in accord with our cherished-belief constraints (items in brackets above) so long as

$$f \gtrsim f_{min} \simeq \left(\frac{c^3/G}{4\pi^2 M_{max} T_H}\right)^{1/2} \simeq 1 \times 10^{-11} \, \text{Hz}; \qquad (9)$$

and thus for these frequencies our cherished beliefs cannot give any limit tighter than (8). Note that $f \gtrsim f_{min}$ includes all frequencies of experimental interest. At lower frequencies, objects of mass $M \lesssim M_{max} = 10^8 \, M_\odot$ radiating with luminosities $L \lesssim (G/c)(2\pi Mf)^2$ cannot radiate away all their mass-energy Mc^2 in a time τ_* less than the age of the Universe T_H; and, consequently, the maximum galactic flux density and wave amplitude are reduced from the limit (8) to

$$\mathscr{F}_f \lesssim \frac{1}{4}\frac{R_g \rho_g c^2}{f T_H}\left(\frac{f}{f_{min}}\right)^2 \simeq 10^{13}\frac{\text{erg}}{\text{cm}^2 \, \text{sec} \, \text{Hz}}\left(\frac{f}{f_{min}}\right), \qquad (10a)$$

$$f^{1/2}\tilde{h} \lesssim 6 \times 10^{-19}(f/1 \, \text{Hz})^{-1}(f/f_{min}) \simeq 6 \times 10^{-8}, \qquad (10b)$$

for

$$f \lesssim f_{min} \simeq 1 \times 10^{-11} \, \text{Hz}.$$

However, this range of frequencies is outside the domain of interest for the present discussion ($10^{-4} \, \text{Hz} \lesssim f \lesssim 10^{+4} \, \text{Hz}$).

IV. Upper Limits on Waves from Discrete Sources

We turn now to gravitational waves from discrete (resolved) sources, including broadband bursts (duration $\tau_* \sim 1/2\pi f$); transient sources ($1/2\pi f \ll \tau_* < \hat{\tau}$, where $\hat{\tau}$ is the total observation time, i.e., the total time that the experimenter searches for gravitational waves); and permanent sources

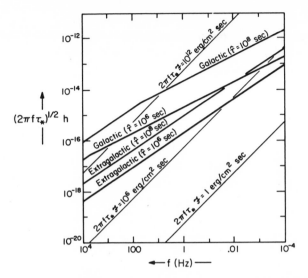

Fig. 2 Upper limits on discrete sources of gravitational waves [Eqs. (11)–(14). These limits answer the following question: "An Experimenter searches, with total observation time τ, for a discrete gravitational-wave event of duration $\tau_* \leq \tau$ at frequencies $f > 1/\tau_* \geq 1/\hat{\tau}$ in a bandwidth $\Delta f \simeq f$. What is the flux $\mathscr{F}$ and amplitude h [Eq. (4) and associated discussion] of the strongest single event he can hope to see, within the constraints of our cherished beliefs?"

($\tau_* \gtrsim \hat{\tau}$). The transient and permanent sources can be either broadband ($\Delta f \sim f$) or narrowband ($\Delta f \ll f$). Our characterization of the waves by their flux $\mathscr{F}$ and amplitude h pays no attention to the bandwidth of the source. Since the experimenter can never know the total "on time" τ_* of the source unless $\tau_* < \hat{\tau}$, and since our cherished beliefs allow stronger waves the shorter is τ_*, we can restrict attention to the case $\tau_* \leq \hat{\tau}$.

For discrete sources our upper limits answer the following question: "An experimenter searches, with total observation $\hat{\tau}$, for a gravitational-wave event of duration $\tau_* \leq \hat{\tau}$ at frequencies $f > (2\pi\tau_*)^{-1}$ in a bandwidth $\Delta f \simeq f$. What is the flux $\mathscr{F}$ and amplitude h of the strongest single event he can hope to see within the constraints of our cherished beliefs?" The upper limits that answer this question are shown in Fig. 2. These limits are derived and discussed, for extragalactic sources, in Section IV.A, and for galactic sources in Section IV.B.

A. Extragalactic Sources

Let the frequency f, event duration τ_*, and observation time $\hat{\tau}$ be given. The waves will be strongest if the bulk of the mass of the Universe resides in sources of some optimally chosen mass M [with $M \lesssim (c^3/G)(2\pi f)^{-1}$], each of which produces some optimal luminosity L during its "on time" τ_* [where $L \lesssim (G/c)(2\pi Mf)^2$], and each of which has some optimal number N of "on events" during the Hubble time T_H [with $NL\tau_* \leq Mc^2$].

The number density of sources is $n \simeq \rho_u/M$, and the probability that a given source will turn on during the observation time $\hat{\tau}$ is $P = N\hat{\tau}/T_H$. Consequently, the nearest source that turns on during $\hat{\tau}$ is at a distance

$$r \simeq \frac{1}{2}(nP)^{-1/3} = \frac{1}{2}\left(\frac{MT_H}{\rho_u N\hat{\tau}}\right)^{1/3};$$

and the flux produced at Earth by this nearest (and thus strongest) source is

$$\mathscr{F} = \frac{L}{4\pi r^2} = \frac{1}{\pi}\left(\frac{N^2 L^3}{M^2}\right)^{1/3}\left(\frac{\rho_u \hat{\tau}}{T_H}\right)^{2/3}.$$

This flux is maximized, subject to our cherished-belief constraints (brackets above) by setting $N = 1$, $M \simeq (c^3/G)(2\pi f)^{-1}$, and $L \simeq Mc^2/\tau_*$ [corresponding to $L \simeq (2\pi f\tau_*)^{-1}(G/c)(2\pi Mf)^2 \lesssim (G/c)(2\pi Mf)^2$]. The resulting upper limit is

$$\mathscr{F} \lesssim \left(\frac{1}{2\pi f\tau_*}\right)\left(\frac{c^3}{\pi G^{1/3}}\right)\left(\frac{\rho_u 2\pi f\hat{\tau}}{T_H}\right)^{2/3}$$

$$\simeq \left(\frac{1 \times 10^7 \text{ erg/cm}^2 \text{ sec}}{2\pi f\tau_*}\right)\left(\frac{f}{1 \text{ Hz}}\right)^{2/3}\left(\frac{\hat{\tau}}{10^6 \text{ sec}}\right)^{2/3}, \qquad (11a)$$

which corresponds to an amplitude

$$h \lesssim \frac{2 \times 10^{-16}}{(2\pi f\tau_*)^{1/2}}\left(\frac{f}{1 \text{ Hz}}\right)^{-2/3}\left(\frac{\hat{\tau}}{10^6 \text{ sec}}\right)^{1/3}. \qquad (11b)$$

The factors $2\pi f\tau_*$ are of order 1 for the most abrupt bursts; slower bursts are constrained to contain the same total energy $Mc^2 = 4\pi r^2 \mathscr{F}\tau_*$ and so produce a lower flux $\mathscr{F} \propto 1/\tau_*$.

B. Galactic Sources

Let the frequency f, event duration τ_*, and observation time $\hat{\tau}$ be given. At sufficiently high frequencies an argument identical to that for extra-galactic sources (Section IV.A) gives the same answer, but with ρ_u replaced by ρ_g: It is optimal for the bulk of the mass of the galaxy to be put into objects of mass $M \simeq (c^3/G)(2\pi f)^{-1}$, which radiate all their mass-energy Mc^2 in single bursts of duration τ_* and luminosity $L = Mc^2/\tau_*$. The strongest burst seen in time $\hat{\tau}$ has flux $\mathscr{F}$ and amplitude h at the upper limit of the inequalities

$$\mathscr{F} \lesssim \left(\frac{1}{2\pi f \tau_*}\right)\left(\frac{c^3}{\pi G^{1/3}}\right)\left(\frac{\rho_g 2\pi f \hat{\tau}}{T_H}\right)^{2/3}$$

$$\simeq \left(\frac{2 \times 10^{10} \text{ erg/cm}^2 \text{ sec}}{2\pi f \tau_*}\right)\left(\frac{f}{1 \text{ Hz}}\right)^{2/3}\left(\frac{\hat{\tau}}{10^6 \text{ sec}}\right)^{2/3}, \qquad (12\text{a})$$

$$h \lesssim \frac{1 \times 10^{-14}}{(2\pi f \tau_*)^{1/2}}\left(\frac{f}{1 \text{ Hz}}\right)^{-2/3}\left(\frac{\hat{\tau}}{10^6 \text{ sec}}\right)^{1/3} \qquad (12\text{b})$$

for $f \gtrsim f_{\text{crit}}$ (defined below).

As one moves to lower and lower frequencies, the optimal scenario corresponds to the strongest event being farther and farther from Earth—at a distance

$$r \simeq \frac{1}{2}\left(\frac{M T_H}{\rho_g \hat{\tau}}\right)^{1/3} = \frac{1}{2}\left[\frac{(c^3/G) T_H}{\rho_g 2\pi f \hat{\tau}}\right]^{1/3}.$$

Ultimately, at critical frequency

$$f_{\text{crit}} \simeq \frac{1}{12}\frac{c^3}{G}\frac{T_H}{M_g \hat{\tau}} \simeq \frac{5 \text{ kHz}}{(\hat{\tau}/10^6 \text{ sec})}, \qquad (13)$$

the distance r has grown to the galactic radius R_g. At frequencies $f < f_{\text{crit}}$, r exceeds R_g and our optimal scenario is no longer valid.

In the low-frequency regime $f < f_{\text{crit}}$, it is optimal to have just one emission event in the entire galaxy during the observation time $\hat{\tau}$, with a mean distance $r \simeq 3R_g/4$ and a luminosity $L \simeq (M_g c^2/\tau_*)(\hat{\tau}/T_H)$ so large that the entire mass of the galaxy will be exhausted in the time T_H. These events correspond to a flux and amplitude at the upper limit of the inequalities

$$\mathscr{F} \lesssim \frac{4}{9\pi}\frac{M_g c^2 \hat{\tau}}{R_g^2 T_H \tau_*} = \left(\frac{2 \times 10^9 \text{ erg/cm}^2 \text{ sec}}{2\pi f \tau_*}\right)\left(\frac{f}{1 \text{ Hz}}\right)\left(\frac{\hat{\tau}}{10^6 \text{ sec}}\right), \qquad (14\text{a})$$

$$h \lesssim \frac{2 \times 10^{-15}}{(2\pi f \tau_*)^{1/2}}\left(\frac{f}{1 \text{ Hz}}\right)^{-1/2}\left(\frac{\hat{\tau}}{10^6 \text{ sec}}\right)^{1/2} \quad \text{for } f \lesssim f_{\text{crit}}. \qquad (14\text{b})$$

Our cherished beliefs permit these events to be produced by objects of mass M anywhere in the range

$$M \lesssim \left(\frac{c^3/G}{2\pi f}\right) \simeq 4M_\odot\left(\frac{f}{f_{\text{crit}}}\right)^{-1}\left(\frac{\hat{\tau}}{10^6 \text{ sec}}\right),$$

$$M \gtrsim \frac{L\tau_*}{c^2} \simeq \frac{M_g \hat{\tau}}{T_H} \simeq 4M_\odot\left(\frac{\hat{\tau}}{10^6 \text{ sec}}\right),$$

$$M \gtrsim \left(\frac{1}{2\pi f}\right)\left(\frac{cL}{G}\right)^{1/2} \simeq \left(\frac{c^3/G}{2\pi f \tau_*}\frac{M_g \hat{\tau}}{2\pi f T_H}\right)^{1/2} \simeq \frac{4M_\odot}{(2\pi f \tau_*)^{1/2}}\left(\frac{f}{f_{\text{crit}}}\right)^{-1/2}\left(\frac{\hat{\tau}}{10^6 \text{ sec}}\right).$$

In this optimal scenario each source must experience $N \simeq Mc^2/L\tau_*$ outbursts in its lifetime. As the frequency decreases far below f_{crit}, it ultimately reaches a limiting value

$$f_{lim} \simeq \left[\frac{(c^3/G)M_g \hat{\tau}}{(2\pi M_{max})^2 \tau_* T_H}\right]^{1/2} \simeq (1 \times 10^{-9}\,\text{Hz})\left(\frac{\hat{\tau}}{\tau_*}\right)^{1/2} \tag{15}$$

at which our optimal scenario requires source masses in excess of $M_{max} = 10^8\,M_\odot$. Below this frequency the flux and amplitude limits (14) are no longer valid; but this ultralow-frequency regime is outside our domain of interest and we shall ignore it.

An attractive (albeit not highly likely) scenario for producing broadband bursts, $\tau_* \sim 1/f$, at kilohertz frequencies, with amplitude h near the upper limit (12b), (14b) is the following: It is fashionable to speculate [25] that before galaxies formed, a sizable fraction of the mass of the Universe may have condensed into massive stars ($M \sim 2$ to $20\,M_\odot$), conventionally called stars of "Population III." A significant fraction of these stars, like stars today, might have formed in close binaries which produce, after the stars have exhausted their nuclear fuel (in $\Delta t \lesssim 1$ billion years), black-hole and/or neutron-star binary systems. When our galaxy condensed out of the intergalactic medium, such binaries would have snuggled down around the galaxy [26] to form a massive halo of the type for which there is strong empirical evidence [9–11]. The orbital parameters of these compact binaries in our halo could perfectly well be such that the mean time for the two stars or holes to spiral together due to gravitational radiation reaction is of order the Hubble time T_H. At the end of its inward spiral, such a binary will emit a sizable fraction of its rest mass (~ 2–20%) in a broadband burst of gravitational waves at kilohertz frequencies [20,27,28]. These bursts could be the events of our optimal galactic scenario.

V. Discussion

It is interesting to compare the cherished-belief upper limits of Figs. 1 and 2 with the sensitivities of gravitational-wave detectors: past, present, and future.

The first-generation Weber-type bars (1968–1976) were capable of detecting broadband bursts occurring once in $\hat{\tau} \simeq 10^6$ sec with frequencies $f \simeq 1000$ Hz and amplitudes $h \gtrsim 3 \times 10^{-16}$. This sensitivity was a little worse than our cherished-belief upper limits (Fig. 2), which explains why theorists could account for Weber's observed events [29] only by invoking unconventional hypotheses (strong beaming by sources near the galactic center [15,16,18]; or today being a very special time in the evolution of the galaxy [30]).

Second-generation detectors of the bar type and laser-interferometer type (1980–1984) are designed to have sensitivities $h \sim 10^{-18}$ for events occurring once in $\hat{\tau} \simeq 10^6$ sec with frequencies $f \simeq 100$–1000 Hz. Such sensitivities are considerably better than our cherished-belief limits (Fig. 2). Thus, although conventional scenarios do not predict waves at this level (sensitivity worse than "type-two benchmarks"), a discovery of waves by second-generation detectors is perfectly possible within the framework of our cherished beliefs.

At much lower frequencies, $f \sim 10^{-3}$ Hz, Doppler tracking of spacecraft is being used to search for gravitational waves. The best sensitivities yet achieved, using the Viking spacecraft [31], correspond to an rms noise level $h_{rms} \sim 3 \times 10^{-14}$ and a sensitivity to $\hat{\tau} = 10^6$ sec bursts of $h \sim 2 \times 10^{-13}$. These sensitivities are slightly worse than our cherished-belief limits. However, future experiments using the Solar Polar spacecraft (1985) and improved tracking technology are projected to have amplitude sensitivities a factor ~ 10 better than Viking's, and a proposed Solar Probe spacecraft (~ 1988) might do a factor ~ 100 better [32]. Such sensitivities would be somewhat better than our cherished-belief upper limits.

In conclusion, the technology of gravitational-wave detection is now crossing over our cherished-belief benchmarks. Near-future experiments will be in a realm where it is not irrational to hope for positive results!

Appendix: Notation

A. Parameters Describing the Structure of the Universe

T_H = Hubble time = 1×10^{10} years = 3×10^{17} sec

$R_H = cT_H$ = Hubble radius = 1×10^{10} lyr = 9×10^{27} cm

ρ_u = mean mass density of Universe = $\dfrac{3c^2/G}{8\pi R_H^2}$ = 1×10^{-8} $M_\odot/\text{lyr}^3$

$\qquad\qquad\qquad\qquad\qquad\qquad\qquad\qquad = 2 \times 10^{-29}$ g/cm^3

R_g = galaxy radius = 6×10^4 lyr = 6×10^{22} cm

M_g = galaxy mass = 1×10^{12} $M_\odot$ = 2×10^{45} g

ρ_g = mean mass density of galaxy = $3M_g/4\pi R_g^3$ = $0.001 M_\odot/\text{lyr}^3$

$\qquad\qquad\qquad\qquad\qquad\qquad\qquad\qquad = 2 \times 10^{-24}$ g/cm^3

$M_{max} = \begin{pmatrix} \text{maximum mass of coherently radiating} \\ \text{object in our galaxy} \end{pmatrix} = 10^8 \, M_\odot$

B. Parameters Describing Gravitational-Wave Sources and Their Radiation

M = mass of coherently radiating source

f = mean frequency emitted by source

L = luminosity of source (erg/sec) in "on" state
τ_* = "on" time for source; burst duration
N = number of "on" events during source's lifetime
n = number density of sources
r = distance to nearest source

C. Parameters Describing Radiation Arriving at Earth

$\hat{\tau}$ = observation time; experiment duration
$\mathscr{F}$ = flux of energy in gravitational waves (erg/cm^2 sec)
h = amplitude of gravitational waves
$\mathscr{F}_f$ = flux density of gravitational-wave background (erg/cm^2 sec Hz)
$\tilde{h}$ = square root of spectral density of amplitude of background radiation (Hz$^{-1/2}$)

Acknowledgment

The question answered by this paper was posed to us by Ronald W. P. Drever and Ranier Weiss. We thank them.

References

[1] Tyson, J. A., and Giffard, R. P., *Annu. Rev. Astron. Astrophys.* **16**, 521 (1978).
[2] Braginsky, V. B., and Rudenko, V. N., *Phys. Rep.* **46**, 165 (1978).
[3] Douglass, D. H., and Braginsky, V. B., *in* "General Relativity: an Einstein Centenary Survey" (S. W. Hawking and W. Israel, eds.), p. 90. Cambridge Univ. Press, London and New York, 1979.
[4] Weiss, R., *in* "Sources of Gravitational Radiation" (L. Smarr, ed.), p. 7. Cambridge Univ. Press, London and New York, 1979.
[5] Thorne, K. S., *in* "Theoretical Principles in Astrophysics and Relativity" (N. R. Lebovitz, W. H. Reid, and P. O. Vandervoort, eds.), p. 149. Univ. of Chicago Press, Chicago, Illinois, 1978.
[6] Epstein, R., and Clark, J. P. A., *in* "Sources of Gravitational Radiation" (L. Smarr, ed.), p. 477. Cambridge Univ. Press, London and New York, 1979.
[7] Grishchuck, L. P., *Pis'ma Zh. Eksp. Teor.* **23**, 326 (1976); Engl. transl. *Sov. Phys.— JETP Lett.* **23**, 293 (1976), and references therein.
[8] Bertotti, B., and Carr, B. J., *Astrophys. J.*, **236**, 1000 (1980); also Carr, B. J., *Astron. Astrophys.*, in press.
[9] Ostriker, J. P., and Peebles, P. J. E., *Astrophys. J.* **186**, 467 (1973).
[10] Bardeen, J., *in* "Dynamics of Stellar Systems" (A. Hayli, ed.), IAU Symposium No. 69, p. 297. Reidel Publ., Dordrecht, Netherlands, 1975.
[11] Toomre, A., *Annu. Rev. Astron. Astrophys.* **15**, 437 (1977).
[12] Oort, J. H., *Annu. Rev. Astron. Astrophys.* **15**, 295 (1977).

[13] Misner, C. M., *in* "Ondes et radiations gravitationelles" (Y. Choquet-Bruhat, ed.), Colloques Internationaux du CNRS No. 220, p. 145. CNRS, Paris, 1974 (and references cited therein).
[14] Kovacs, S. J., Jr., and Thorne, K. S., *Astrophys. J.* **224**, 62 (1978).
[15] Lawrence, J. K., *Astrophys. J.* **171**, 483 (1971).
[16] Jackson, J. C., *Nature (London)* **241**, 513 (1973).
[17] Press, W. H., *Phys. Rev. D* **15**, 965 (1977).
[18] Misner, C. W., *Phys. Rev. Lett.* **28**, 994 (1972).
[19] Davis, M., Ruffini, R., Press, W. H., and Price, R. H., *Phys. Rev. Lett.* **27**, 1466 (1971).
[20] Detweiler, S. L., and Szedenits, E., Jr., *Astrophys. J.* **231**, 211 (1979).
[21] Cunningham, C. T., Price, R. H., and Moncrief, V., *Astrophys. J.* **224**, 643 (1978); *Astrophys. J.* **230**, 870 (1979).
[22] Einstein, A., *Berl. Sitzungsber.* p. 154 (1918).
[23] Dyson, F. J., *in* "Interstellar Communication" (A. G. W. Cameron, ed.), p. 115. Benjamin, New York, 1963.
[24] Rees, M., *in* "Ondes et radiations gravitationelles" (Y. Choquet-Bruhat, ed.), Colloques Internationaux du CNRS No. 220, p. 203. CNRS, Paris, 1974.
[25] Truran, J. W., and Cameron, A. G. W., *Astrophys. Space Sci.* **14**, 179 (1971).
[26] Gunn, J. E., *Astrophys. J.* **218**, 592 (1977).
[27] Clark, J. P. A., van den Heuvel, E. P. J., and Sutantyo, W., *Astron. Astrophys.* **72**, 120 (1979).
[28] Detweiler, S. L., *Astrophys. J.* **225**, 687 (1978).
[29] Weber, J., *Phys. Rev. Lett.* **22**, 1302 (1969); **24**, 6 (1970).
[30] Kafka, P., *Nature (London)* **226**, 436 (1970).
[31] Armstrong, J. W., Woo, R., and Estabrook, F. B., *Astrophys' J.* **230**, 570 (1979).
[32] Estabrook, F. B., *in* "A Close-Up of the Sun" (M. Neugebauer and R. W. Davies, eds.), JPL 78-70, p. 441. Jet. Propul. Lab., Pasadena, California, 1978.

11

General Relativistic Hydrodynamics: The Comoving, Eulerian, and Velocity Potential Formalisms

Larry Smarr and Clifford Taubes

Harvard-Smithsonian Center for Astrophysics

and

Lyman Laboratory of Physics
Harvard University
Cambridge, Massachusetts

and

James R. Wilson

Lawrence Livermore Laboratory
University of California
Livermore, California

Abstract

We compare the three major approaches to general relativistic hydrodynamics: comoving, Eulerian, or mixed Euler–Lagrange, and velocity potential. We show that the comoving gauge has three major disadvantages: (1) shocks cause discontinuities in the spacetime metric; (2) the formation of black holes can lead to time slicing singularities; (3) coordinate lines can become tangled in nonspherical flows. The mixed Euler–Lagrange scheme we use solves these three problems. We show how the comoving limit is obtained from our general equations. The circulation of a fluid flow in the presence of shocks is discussed and illustrated by numerical example. We suggest a possible relevance for this to supernova ejection mechanisms. Finally, the velocity potential formalism is discussed and compared to the comoving and Eulerian approaches. Taub's contributions are stressed throughout the discussion.

157

I. Introduction

This paper arose from a series of discussions with Abe Taub over the last few years. Having been one of the strongest proponents of the use of comoving coordinates to describe relativistic hydrodynamics, he was interested in the possible limitations of this approach. We discuss that question here by using the ADM [1] kinematic analysis of spacetime coordinate systems [2]. We find three significant difficulties with the comoving gauge. First, shocks will cause discontinuities in either the lapse function or the shift vector. Second, if the comoving gauge is used to specify the time slicing, as in May and White's [3,28] or Pachner's [4] work, then the time slicing will become singular shortly ($\sim GMc^{-3}$) after the formation of a black hole. Third, if one is following nonspherical flow, then the growth of circulation behind shocks will tangle the spatial coordinates.

We show how removing the comoving restriction solves these three problems at the expense of having to integrate the full Eulerian hydrodynamics equations. The comoving limit is easily obtained from our equations. In the discussion of the third problem listed above, we review the circulation theorem in relativistic hydrodynamics following Taub's classic account [5]. A numerical example using the Eulerian approach shows clearly the generation of circulation after shocks form in a nonspherical collapse. We briefly discuss the relevance of this process for supernovae following Epstein [6] and Colgate's [7] recent remarks.

Finally, we review the relativistic velocity potential formalism introduced by Schutz [8]. We show how it related to both the comoving and Eulerian schemes and discuss its possible utility for numerical hydrodynamics.

II. Kinematics of Flows on Spacetime

As described in detail elsewhere [2], one can consider the coordinate system that labels spacetime events as a pair of flows on spacetime. These flows are smooth congruences of curves or trajectories. One flow is normal to the chosen foliation of time slices ($t = $ constant). A unit tangent vector to this flow $\hat{n}^a$ defines the Eulerian observers in this coordinate system. The term Eulerian refers to the fact that these observers (unit timelike vectors) are at rest in the chosen time slices. The other flow is the spatial coordinate trajectories along which x^i ($i = 1, 2, 3$) are constant. The unit tangent vector for this flow is denoted $\hat{t}^a$ and it can be either timelike, null, or spacelike. The spacetime metric tensor g_{ab}, written in this coordinate system [1,2], defines the lapse function $\alpha \equiv (g^{tt})^{-1/2}$, the shift vector $\beta_i \equiv g_{ti}$, and the three metric $\gamma_{ij} \equiv g_{ij}$.

In addition to these two flows, if the spacetime is not vacuum, there is the matter flow with unit timelike vector $\hat{u}^a$. The trajectories of this flow will not coincide with those of $\hat{n}^a$ or $\hat{t}^a$ in general. Thus, at each event in spacetime one has three unit vectors $(\hat{n}^a, \hat{t}^a, \hat{u}^a)$ that can be related by boosts. (We assume for now that $\hat{t}^a$ remains timelike; for the spacelike case, see Smarr and Wilson [9].) Since $\hat{n}^a$ is the Eulerian observer, it is natural to refer the boosts to it. In the coordinate system (t, x^i) defined by (α, β^i) above, these vectors are†

$$\hat{n}_a = (-\alpha, 0), \qquad\qquad \hat{n}^a = (\alpha^{-1}, -\beta^i/\alpha), \qquad (1a)$$

$$\hat{t}_a = \Gamma_1(-\alpha + \beta_k\beta^k/\alpha, \beta_i/\alpha), \qquad \hat{t}^a = \Gamma_1(\alpha^{-1}, 0), \qquad (1b)$$

$$\hat{u}_a = \Gamma_2(-\alpha + \beta_k v^k, v_i), \qquad \hat{u}^a = \Gamma_2(\alpha^{-1}, v^i - \beta^i/\alpha), \qquad (1c)$$

$$-n \cdot t \equiv \Gamma_1 = (1 - \beta_k\beta^k/\alpha^2)^{-1/2}, \qquad -n \cdot u \equiv \Gamma_2 = (1 - v_k v^k)^{-1/2}. \qquad (1d)$$

Thus, the boost velocity of $\hat{u}^a$ relative to $\hat{n}^a$ is v^i, while that of $\hat{t}^a$ relative to $\hat{n}^a$ is β^i/α. The two relativistic gamma factors are Γ_1 and Γ_2.

We can decompose the spacetime covariant derivative of each of these unit vector fields into its twist ω^{ab}, shear σ^{ab}, expansion θ, and acceleration ζ^a (for definitions, see Smarr and York Eqs. (4.2) and (4.3) [2]):

$$\nabla^a\hat{Z}^b = \omega^{ab} + \sigma^{ab} + \tfrac{1}{3}h^{ab}\theta - \hat{Z}^a\zeta^b. \qquad (2a)$$

Here $\hat{Z}^b$ is an arbitrary unit timelike vector with its projection operator $\hat{h}^{ab}$. For the Eulerian observers $\hat{n}^a$, we have

$$\omega^{ab} = 0, \qquad (2b)$$

$$\theta^{ab} = \sigma^{ab} + \tfrac{1}{3}\theta h^{ab} = \begin{pmatrix} 0 & 0 \\ 0 & -K^{ij} \end{pmatrix}, \qquad (2c)$$

$$h^{ab} = \begin{pmatrix} 0 & 0 \\ 0 & \gamma^{ij} \end{pmatrix}, \qquad (2d)$$

$$\zeta^a = (0, \gamma^{ij}\partial_j \ln \alpha), \qquad (2e)$$

where K^{ij} is the extrinsic curvature tensor of the time slice. We see that if the time slices are not geodesically parallel ($\alpha = 1$), then *the Eulerian observers are accelerated* and not inertial as in flat spacetime.

Continuing the fluid analogy, one can write down the propagation equations along $\hat{n}^a$ for the above quantities [10]:

$$\mathscr{L}_n h_{ab} = 2\theta_{ab}, \qquad (3a)$$

$$\mathscr{L}_n \theta_{ab} = \theta^c_a\theta_{cb} + n^c R_{cabd}n^d + D_a\zeta_b + \zeta_a\zeta_b, \qquad (3b)$$

† Latin letters from the first part of the alphabet range over 0,1,2,3; those from the middle part of the alphabet range over 1,2,3. We also set $G = c = 1$.

where $\mathscr{L}_n$ is the Lie derivative along $\hat{n}^a$ and R_{cabd} is the Riemann tensor of g_{ab}. With the slicing (α) fixed, one can then use the identity

$$\mathscr{L}_{\alpha n} = \alpha\mathscr{L}_n = \mathscr{L}_t - \mathscr{L}_\beta, \tag{4}$$

where

$$t^a = \alpha\hat{n}^a + \beta^a = (\alpha^2 - \beta^2)^{-1/2}\hat{t}^a \tag{5}$$

and $\beta^a = (0, \beta^i)$. This, together with using the Einstein field equations $G_{ab} = 8\pi T_{ab}$ to evaluate the four-dimensional R_{abcd} in terms of three-dimensional quantities, yields the standard ADM form of the Einstein field equations [1,2]:

$$\mathscr{L}_t\gamma_{ij} - \mathscr{L}_\beta\gamma_{ij} = -2\alpha K_{ij}, \tag{6a}$$

$$\mathscr{L}_t K_{ij} - \mathscr{L}_\beta K_{ij} = -\alpha\mathscr{K}_{ij} + \alpha\mathscr{R}_{ij} - D_i D_j\alpha - 8\pi\alpha\mathscr{S}_{ij}, \tag{6b}$$

where we have grouped terms to parallel Eq. (3) by defining [2]:

$$\mathscr{K}_{ij} \equiv 2K_{mi}K_j^m - (\text{tr } K)K_{ij} + \tfrac{1}{4}[(\text{tr } K)^2 - K_{mn}K^{mn}]\gamma_{ij}, \tag{6c}$$

$$\mathscr{R}_{ij} \equiv R_{ij} - \tfrac{1}{4}\text{tr}(R)\gamma_{ij}, \tag{6d}$$

$$\mathscr{S}_{ij} \equiv S_{ij} - \tfrac{1}{2}\text{tr}(S)\gamma_{ij}. \tag{6e}$$

Here S_{ij} is the stress of the energy–momentum tensor T_{ab} (see below). It is clear that *the Einstein field Eqs. (6) can be regarded as propagation equations for the shear and expansion of the Eulerian observer congruence*. In fact, the expansion $\theta = -\text{tr}(K)$ obeys the trace of the field equations (6):

$$\mathscr{L}_t(\text{tr } K) - \mathscr{L}_\beta(\text{tr } K) = -\Delta\alpha + \tfrac{1}{4}\alpha[\text{tr}(R) + (\text{tr } K)^2 + 3K_{mn}K^{mn} + 16\pi\,\text{tr}(S)], \tag{7}$$

where Δ is the three-covariant Laplacian formed from the three-covariant derivative D_i. Since the specification of the lapse function α defines the time slicing, we see that *the expansion of the Eulerian observers congruence is determined by the time slicing coordinate condition*. Finally, it is often useful to use the four constraint equations of general relativity:

$$D^m[K_{mi} - \gamma_{mi}\,\text{tr}(K)] = 8\pi j_i, \tag{8a}$$

$$\text{tr}(R) + [\text{tr}(K)]^2 - K_{mn}K^{mn} = 16\pi d, \tag{8b}$$

where j_i and d are defined from T_{ab} in Section III.

III. Hydrodynamics and Thermodynamics

We briefly review here the behavior of a perfect fluid, characterized by a rest-mass density ρ, a specific internal energy ε, and an isotropic pressure P. These are defined in the rest frame of the fluid; we may write the energy–momentum–stress tensor as

$$T_{ab} = \rho h \hat{u}_a \hat{u}_b + P g_{ab}, \tag{9}$$

where the quantity

$$h = 1 + \varepsilon + P/\rho \tag{10}$$

is the specific enthalpy. Because of the lack of standardized notation in the literature we provide in Table I a dictionary that compares our notation with that of several standard works. The quantity h defined in Eq. (10) is an essential quantity in relativistic hydrodynamics. Besides generalizing the notion of specific enthalpy, ρh plays the role of the "effective inertial mass" of the fluid [Eq. (19)]. It can also be considered as the injection energy or chemical potential, an interpretation much used by Thorne [11]. Finally, we show that in the comoving gauge, it is h that determines the kinematics.

Since the Einstein equations (6, 8) are tied to the Eulerian observers, we must boost T_{ab} from the $\hat{u}^a$ frame to the $\hat{n}^a$ frame. To the Eulerian observer the perfect fluid appears imperfect [compare (9) above]:

$$T_{ab} = d\hat{n}_a \hat{n}_b + j_a \hat{n}_b + j_b \hat{n}_a + S_{ab}. \tag{11}$$

That is, the fluid appears to have momentum flux and anisotropic shear stresses. By taking the appropriate contraction of (9) with $\hat{n}^a$ and using Eq. (1d), one can easily calculate (d, j_i, s_{ij}):

$$d = \rho h (\Gamma_2)^2 - P, \tag{12a}$$

$$j_i = -\rho h (\Gamma_2)^2 v_i, \tag{12b}$$

$$S_{ij} = P\gamma_{ij} + \rho h (\Gamma_2)^2 v_i v_j, \tag{12c}$$

which agrees, for instance, with Eq. (125.3) in Landau and Lifshitz [12]. These quantities act as the source terms for Einstein's equations (6, 8).

The basic fluid variables (ρ, ε, P) are augmented by the temperature T and the specific entropy S. The first law of thermodynamics relates these quantities:

$$T dS = d\varepsilon + P d(1/\rho). \tag{13}$$

An alternate form that will prove useful is

$$T dS = dh - (1/\rho) dP. \tag{14}$$

TABLE I

Hydrodynamics notation

Quantity/Reference	Landau and Lifshitz [12]	Taub [14]	Misner and Sharp [24]	May and White [3]	Schutz [8]	Lichnerowicz [50]	Carter [46]
Rest-mass density ρ	nmc^2	ρc^2	n	ρ	ρ_0	r	ρ
Mass-energy density $\rho(1+\varepsilon)$	e	$w = \rho(c^2 + \varepsilon)$	ε	$\rho(c^2 + \varepsilon)$	$\rho = \rho_0(1+\Pi)$	$\rho = r(1+\varepsilon)$	ρ
Enthalpy $h = 1 + \varepsilon + P/\rho$	w/nmc^2	$c^2 + i$	$h = (\varepsilon + P)/n$	$1 + \dfrac{\varepsilon}{c^2} + \dfrac{P}{\rho c^2}$	$1 + \Pi + P/\rho_0$	$f = 1 + i$	$\mu = \dfrac{\rho + P}{n}$
Specific entropy S	σ/nmc^2	S	s	—	S	S	s
Momentum $S_i/(\rho f)$	wv_i	$V^\mu = \left(1 + \dfrac{i}{c^2}\right)u^\mu$	—	—	$V_\sigma = \mu u_\sigma$	$C_\alpha = f u_\alpha$	$\pi_\alpha = \mu u_\alpha$
Constants $G = c = 1$	G, c	G, c	$G = c = 1$	G, c	$G = c = 1$	G, c	$c = 1$
Signature	+	−	+	−	+	−	+

Much of the literature on relativistic hydrodynamics considers only *isentropic* flow $dS = 0$. Our discussion will not make this restriction, but we shall point out how equations simplify in that case.

One must specify an equation of state that defines one function $(\rho, \varepsilon, P, S)$ in terms of two others, e.g., $P = P(\rho, S)$. Often, discussions of relativistic hydrodynamics assume a *barotropic* equation of state in which the dependent variable is a function of only one quantity, e.g., $P = P(\rho)$. A special example is the isentropic case, $S = $ constant, mentioned above. This restriction is inappropriate to the physical situation of relativistic collapse when shocks occur. (For a comparison of the barotropic and nonbarotropic equations, see Taub [13].)

The conservation laws for energy and for momentum are contained in the four equations:

$$\nabla^a T_{ab} = 0. \tag{15}$$

Just as the form of T_{ab} takes a simple form in the rest frame of the fluid (defined by $\hat{u}^a$), so do these equations. Following Taub's derivation [14] we first contract (15) with $\hat{u}^b$ and then use the first law of thermodynamics (13a) to obtain

$$\nabla_a(\rho\hat{u}^a) + \rho h^{-1} T \hat{u}^a \nabla_a S = 0. \tag{16}$$

When this is combined with the law of baryon conservation

$$\nabla_a(\rho\hat{u}^a) = 0, \tag{17}$$

we derive that the perfect fluid is *locally adiabatic*:

$$\hat{u}^a \nabla_a S = 0. \tag{18}$$

That is, entropy is constant along the flow lines of the fluid. Note that if the fluid had been assumed to be isentropic ($\nabla_a S = 0$), we would not need to assume baryon conservation as an independent law, but rather would have derived it from (15). The other three equations are obtained by projecting (15) into the rest frame of $\hat{u}^a$ using $\tilde{h}_{ab} \equiv g_{ab} + \hat{u}_a \hat{u}_b$:

$$\rho h \, \hat{u}^a \nabla_a \hat{u}_b = \tilde{h}^a_b \nabla_a P. \tag{19}$$

This is just the Euler equation stating that the specific effective mass (ρh) times the fluid's acceleration $(\hat{u}^a \nabla_a \hat{u}_b)$ is the pressure gradient, the fluid version of Newton's law of motion.

In summary, the eight unknown fluid quantities $(\rho, \varepsilon, P, S, \hat{u}_a)$ are determined by Eqs. (17), (13), equation of state (18), (19), and the restriction $\hat{u}^a \hat{u}_a = -1$. We apologize to the reader for repeating well-known material, but it will be useful to have set notation and to have made the distinctions between adiabatic, isentropic, and barotropic.

A final cautionary note: In the real world, fluids shock. Across these discontinuities the entropy of fluid worldlines change, contrary to the law (18). This is because a fluid is dissipative in a shock front and thus imperfect even in its rest frame, violating the form (9) used to derive (18). Thus, one should think of the fluid as distinct regions, in which the above equations hold, separated by shock fronts, across which jump conditions replace (15) and (17). (For detailed discussion of these jump conditions, see Taub [15] and Thorne [16].)

IV. Lagrangian or Comoving Coordinates

One choice of spacetime coordinates is to force the spatial coordinate trajectories to coincide with the matter flow. In fact, until the work of Wilson [17], all relativistic hydrodynamics used this gauge. Taub [13] has provided a comprehensive review of this choice, which we draw from here. Our purpose is twofold. First, we wish to show how the particular calculations in comoving coordinates made by many authors [3,4,22–34] are all special cases of Taub's general comoving equations. Second, we wish to show how Taub's equations are specializations of the mixed Euler–Lagrange system we use. We divide this section into three parts. The general comoving equations, the spherical case, and the nonspherical case.

A. The Comoving Gauge

The requirement that $\hat{t}^a = \hat{u}^a$ to guarantee comoving coordinates is satisfied if $(1b) = (1c)$ or

$$v^i = \beta^i/\alpha, \tag{20a}$$

$$\Gamma_1 = \Gamma_2 \equiv \Gamma = (1 - \beta_k \beta^k/\alpha^2)^{-1/2}, \tag{20b}$$

$$\hat{u}^a = \hat{t}^a = (\alpha^2 - \beta_k \beta^k)^{-1/2}(1, 0), \tag{20c}$$

$$\hat{u}_a = \hat{t}_a = (\alpha^2 - \beta_k \beta^k)^{-1/2}(-\alpha^2 + \beta_k \beta^k, \beta_i). \tag{20d}$$

These are Taub's Eqs. (15.1) and (15.2) [13] in our notation. By substituting them into the Euler equation (19) and using the first law (14), Taub obtains the comoving gauge equation, his (15.4):

$$\partial_t[h\beta_i(\alpha^2 - \beta_k \beta^k)^{-1/2}] + \partial_i[h(\alpha^2 - \beta_k \beta^k)^{1/2}]$$
$$+ T(\alpha^2 - \beta_k \beta^k)^{1/2}\partial_i S = 0 \tag{21}$$

This will be the master equation for unraveling the many special choices of comoving gauge used by other authors. We note that Taub was not the first to derive Eq. (21). It goes back at least to Tauber and Weinberg [18], see their Eqs. (127) and (128). Tolman [19] states that one can always find a comoving coordinate system, so the idea dates from at least the 1930s.

Assuming the fluid quantities (h, T, S) are known, the three equations (21) are conditions on the gauge variables (α, β_i). One important drawback of the comoving gauge was noted by Taub [14,20]. From (21), one can see that if shocks form, with discontinuities in (h, T, S), these *must* cause discontinuities in (α, β_i). Thus, the comoving coordinate system will not be a smooth one on spacetime. This major defect will be illustrated below.

The great advantage of the comoving gauge is that the equations of hydrodynamics are automatically integrated. Already we have solved for $\hat{u}^a$ in terms of (α, β_i). Since from (18) and (20c), $\hat{u}^a \nabla_a S = (\alpha^2 - \beta_k \beta^k)^{-1/2} \partial_t S = 0$, the entropy is constant along each spatial coordinate trajectory:

$$S = S(x^i), \tag{22}$$

except where these trajectories cross shock fronts. The rest-mass density (or number density) is easily found from (17) and (20c) (Taub [13, Eq. 15.3])†:

$$\rho = \frac{(\alpha^2 - \beta_k \beta^k)^{1/2} f(x^i)}{\alpha [\det(\gamma)]^{1/2}}. \tag{23}$$

Thus, ρ is determined by its initial value and the metric. Having used up the conservation equations (15) and (17), we are left with the first law (13a), which reads as a result of (22):

$$\partial_t \varepsilon = -P \, \partial_t (1/\rho), \tag{24}$$

where $P = P(\rho, S)$ completes the solution of the fluid variables.

If α is fixed by some time slicing condition and β_i is determined by (21), this leaves only γ_{ij} to solve for. This can conveniently be done by using Einstein's equations in the ADM form (6, 8) with source terms given by (12). Taub [21] writes down these equations as his (2.5, 2.6, 2.9). Note that some metric functions may be evolved (6) and some solved for by constraint equations (8).

The only remaining question is that of choosing a time function. We have left α free so that it might be chosen in a manner independent of the matter flow, e.g., by maximal slicing (see below). However, Taub [13] has shown that one can choose a time slicing that depends on the flow and in a sense complements the comoving spatial trajectories. We interpret his choice as follows. The matter flow has a part due to radial flow and a part due to circulation. As we show below, the circulation has contributions from both

† Taub's notation differs from ours as follows: his → ours; $k^{\mu\nu} \to \gamma^{ab}$; $g_{ij} \to \gamma_{ij}$; $N \to \alpha$; $p_i \to \beta_i$; $K_{ij} \to -K_{ij}$; $K \to -K$; $e^\phi \to (\alpha^2 - \beta_k \beta^k)^{1/2}$; $v^i \to \beta^i/(\alpha^2 - \beta_k \beta^k)$.

entropy gradients and vorticity. Since the radial flow is hypersurface ortho-gonal, one can bend the time slice so this part of the flow is at rest in the time slice. This determines α. The remainder of the flow (circulation) is made comoving by use of β_i. From Eq. (15.8) in Taub [13], we find that in our language:

$$\alpha = h^{-1}[1 + h^{-2}(C_k + \theta S_{,k})(C^k + \theta S^{,k})]^{1/2} \tag{25a}$$

and

$$\beta_i = -h^{-2}(C_i + \theta S_{,i}) \tag{25b}$$

define Taub's special comoving frame. Note that both $C_k(x^i)$ and $S_{,k}(x^i)$ are determined by initial conditions, with C_k specifying any initial vorticity. To see this one can calculate the twist (2a) of $\hat{u}^a$ in these coordinates. (See, e.g., Tauber and Weinberg [18], MacCallum and Taub [21], and Pachner [4].)

B. Spherical Spacetimes in Comoving Gauge

Most general relativistic hydrodynamical calculations have been spherical and used comoving coordinates [3,22–34,38,40]. However, these calculations have always derived their equations after the assumption of spherical symmetry. Here we show how they follow from the general comoving equations valid for multidimensional flow.

The first important calculations were those of Tolman [22] and Bondi [23] where the spherical flow was assumed to be pressureless. Since here $p = \varepsilon = S = C_k = 0, h = 1$ and Taub's special comoving frame (25) becomes

$$\alpha = 1, \qquad \beta_i = 0, \tag{26}$$

which, of course, is the geodesic normal coordinates that are natural for dust spacetimes. As is well known, such time slices become singular as the dust free-falls to infinite density {from Eq. (23) $\rho = [\det(\gamma)]^{-1/2}$}. The three-metric has two functions ($\gamma_{rr}, \gamma_{\theta\theta} = \gamma_{\phi\phi} \sin^{-2} \theta$). One, γ_{rr}, can be integrated by the constraint equation (8b) to give the mass function [Bondi, Eq. (5)], while the other is evolved using (6). These latter equations can be written analogously to Newton's law of motion [Bondi, Eq. (34)].

In the 1960s there was a burst of activity extending the dust calculations to include pressure. The spherical comoving equations were worked out independently by several authors [3,24–28]. These equations were finite differenced and solved on the computer for supernova collapse [3,27–31] and supermassive star collapse [32–34]. We shall not give a detailed com-parison of the methods here since they are all equivalent as far as their kinematics are concerned. We particularly study May and White's [3,28] calculations because they are extensively documented.

Misner and Sharp [24] added an isentropic pressure. One now has only $C_k = S_{,k} = 0$, so the Taub gauge (25) becomes

$$\alpha = h^{-1}, \qquad \beta_i = 0, \tag{27}$$

which is Misner and Sharp's equation (1.5). Note that as the collapse proceeds, α becomes less than one in the central region since $\varepsilon \geq 0$ and $P \geq 0$. May and White [3,28] extended this by dropping the restriction that the equation of state be barotropic. As a result, only $C_k = 0$ and Taub's special gauge is

$$\alpha = h^{-1}[1 + h^{-2}T^2 S_{,k} S_{,}^k]^{-1/2}, \tag{28a}$$

$$\beta_i = -h^{-2} TS_{,i}. \tag{28b}$$

This is *not* the gauge used by May and White since they choose $\beta_i = 0$. They apparently did so because of a remark of Tolman's [19] that in spherical comoving coordinates, one can always set $\beta_i = 0$. This is true, but it places severe restrictions on the region of space–time covered by the coordinate system. To obtain their equations one can use the comoving gauge equation (21), which when $\beta_i = 0$, becomes the lapse equation:

$$\partial_i(\alpha h) + T\alpha S_{,i} = 0. \tag{29}$$

With the use of the first law (14) this is Eq. (26) in May and White [28]. Note when $S_{,i} = 0$, this reduces to Misner and Sharp's gauge (27). We examine the consequences of this gauge choice below.

The fluid equations of May and White can be obtained directly from the general comoving equations discussed in Section IV.A. Two are identical: our (22) and (24) are May and White's [28] (27). Our (23) becomes their (22) on noting that in their metric $[\det(\gamma)]^{1/2} = bR^2$, $(\gamma_{\mu\mu} \equiv b^2, \gamma_{\theta\theta} \equiv R^2, \mu$ is the radial coordinate). As in the dust case of Tolman [22] and Bondi [23], the radial metric function is redefined in terms of $m(r, t)$, which is then solved for May and White's Eq. (38) by the Hamiltonian constraint equation (8b). The two-sphere metric function R^2 and its extrinsic curvature $K_{\theta\theta} = U/R$ are evolved by May and White's Eqs. (33) and (34) [28], which are derived from our (6) combined with (8) (see Misner and Sharp [24], Section VII for derivation).

May and White's beautiful pair of papers [3,28] can be interpreted to yield some general lessons about the use of the comoving gauge. There are two separate failures of this gauge. First, when a shock forms, the lapse function α (or a in May and White's notation) becomes discontinuous.† May and

---

† V. Moncrief (personal communication) has pointed out that a discontinuous lapse does *not* imply that the time slice itself is discontinuous in spacetime, but rather it is only C^0 across the timelike shock front.

White calculate that across the shock $[ah] = 0$ so that the jump in the lapse a just compensates the jump in the enthalpy h. Even when the shock is in a weak gravitational field, the lapse is still discontinuous (see Eq. (82) in May and White [28]). Their numerical calculation of a neutron star collapse and bounce (May and White [3], Fig. 5) clearly shows this. Note the jump in a is *not* a small effect, but of order the values $1 - \alpha_c$ for the central lapse α_c.

The second failure occurs where continued gravitational contraction leads to the formation of a black hole. In this case the time slicing defined by (29) becomes singular shortly after the black hole forms (May and White [3], Figs. 8–10). The question of which time slicing conditions will avoid singularities has recently been extensively investigated [2,35,36]. As shown by Smarr and York [2], the lapse must fall toward zero in the strong field region sufficiently rapidly that the integrated proper time

$$\tau_p = \int_{t_0}^{t} \alpha(x^i, t')dt' \tag{30}$$

is less than the proper time from the hypersurface $t = t_0$ to the singularity. For instance, Smarr and York showed that $\alpha \sim e^{-t/\tau_0}$ for maximal time slicing (see below) so that τ_p is finite and at least has a chance of avoiding the singularity. (For cases where it fails, see Smarr and Eardley [35].) May and White observed that $\alpha \to 0$ in their gauge as well ([3], Fig. 11), but calculated that it did so only as a power law [3]:

$$a \sim t^{-3(\gamma - 1)} \tag{31}$$

for equation of state

$$P = (\gamma - 1)\rho\varepsilon. \tag{32}$$

See May and White [3], Eqs. (30) and (31), noting the misprint that left a t out of the first line of their (31). Thus for a typical collapse to a black hole $\gamma \sim \frac{4}{3}$ and $a \sim t^{-1}$. In this case τ_p *diverges* with t as $\ln t$, so we conclude that *May and White's time gauge* (29) *cannot possibly avoid the singularity*. In addition, the time slice first hits the singularity at $\mu \sim 8M_\odot$ for the $21M_\odot$ collapse studied by May and White, enclosing the central region in a "bag of gold" singularity [37]. The time between when the first trapped surface appears on a time slice ($t \sim 0.6644$, Fig. 17) and when $R \to 0$ ($t \sim 0.6645$, Fig. 9) is $\sim 10^{-4}$ sec or $\sim GMc^{-3}$ for a $21M_\odot$ object as expected.

This second failure can be remedied by choosing a different time slicing (α) and using a comoving shift vector determined by (21). Chrzanowski [38] modified May and White's code to do this. If maximal slicing [39] is used to fix α, then the time slicing will go inside any black holes formed, but will usually [35] remain nonsingular for arbitrarily large times after the black hole is formed. This is particularly important if one wishes to study accretion onto black holes. Of course, if the matter shocks, a discontinuity will develop in β^i and the first failure of comoving coordinates mentioned above will still occur.

We mention in passing that May and White's code can be applied to nonasymptotically flat space-times. Novikov and Polnarev [40] have recently used a similar code to study black hole formation in the early Universe. The two failures mentioned above occur here as well. Another variation on May and White is to include a transport equation. First, Misner and Sharp [25] and Schwartz [29] added a radiative diffusion term. Wilson [30] improved this by using a general relativistic Boltzmann neutrino transport equation [41] to see if neutrinos could blow off the envelope of a collapsing star.

C. Nonspherical Spacetime in Comoving Gauge

When one drops spherical symmetry, one must also drop Tolman's dictum that the shift vector β_i can always be made zero. This is because non-spherical flows can exhibit vorticity, which is nonhypersurface orthogonal. That is, there does not exist a global time slice in which the matter is at rest. The idea of using a shift vector $\beta_i = g_{0i}$ such that coordinates comove with the rotating matter was extensively treated by Tauber and Weinberg [18].

Pachner [4] has used the comoving formalism to try to calculate rotating stellar collapse. His series of papers [4] revived the comoving ideas at about the same time Taub was using them for his variational principles. These works seem to have been independent. Pachner [4] restricts the comoving gauge equation (21) to the isentropic case ($\partial_i S = 0$) by his equation $d\varepsilon/d\rho = P\rho^{-2}$. He then uses it to fix his gauge by breaking (21) into two independent pieces, each of which is set equal to zero: the first term of (21) determines β_i and the second term determines $g_{00} = -\alpha^2 + \beta_i \beta^i$. (The third term vanishes because of isentropic flows.) This leads to the isentropic case of Taub's special comoving frame (25). (Pachner's $e^\chi = h$, $A_i = C_i$, and $A_4 = 1$, in our notation.)

Thus, Pachner's gauge can be seen to be the rotating generalization of Misner and Sharp's spherical gauge (27). As such, it will suffer the same two failures of comoving coordinates discussed above (assuming Pachner dropped his isentropic restriction so shocks could be allowed). However, as we shall see below, the nonspherical comoving gauge has in addition a third failure. When shocks form, the resulting entropy gradient drives circulation and thus vorticity. These tend to characteristically form swirls (see Figs. 1–3). If the spatial coordinates attempt to follow this motion (required by the comoving assumption), they will become badly snarled and perhaps self-intersect. It seems likely that this lack of smoothness in the spacetime coordinate system will be at least as damaging as the discontinuities that will develop at the shocks themselves.

V. Eulerian or Noncomoving Coordinates

We have shown in the previous section that the requirement that the space-time coordinate system exactly follow the matter flow on spacetime leads to three distinct failures of the smoothness of the coordinates. First, shocks cause discontinuities in (α, β_i). Second, time slices chosen by the matter flow (Tolman–Bondi, Misner–Sharp, May–White, Pachner) become singular when black holes form. Third, in nonspherical flow shocks develop swirling matter flow behind them that tangles coordinate lines. In exchange for this, one can easily integrate the fluid variables by (20), (22)–(24).

From a numerical relativity point of view, it would seem to be very important to retain smooth spacetime coordinates in which to describe the gravitational field. This is particularly important when gravitational waves are generated by the mass motion and must propagate outward through the coordinates. Therefore, we have given up the use of the gauge freedom (α, β_i) to simplify the hydrodynamics equations and instead have used it to simplify the description of the gravitational field itself. The price we pay is that we must now integrate the full Eulerian hydrodynamics equations. However, we would have to do that anyway for multidimensional flows to avoid the third failure of the Lagrangian gauge mentioned above. Therefore, we now describe the equations we use and show how they reduce to the comoving ones in the appropriate limit.

A. Eulerian Hydrodynamics

In order to write down the equations of hydrodynamics in the ADM coordinate system defined by (α, β_i), we use the representation (1). It will be useful to define some intermediate variables [17]:

$$D \equiv \rho \hat{u}^t = \rho \Gamma_2 \alpha^{-1}, \tag{33a}$$

$$E \equiv \varepsilon \rho \hat{u}^t = \varepsilon \rho \Gamma_2 \alpha^{-1}, \tag{33b}$$

$$\hat{u}^a = V^a \hat{u}^t, \qquad V^a = [1, \alpha v^i - \beta^i], \tag{33c}$$

$$S^a = \rho h \hat{u}^a \hat{u}^t = \rho h (\Gamma_2 \alpha^{-1})^2 (1, V^i). \tag{33d}$$

We shall treat (D, E, S^i) as our mass density, internal energy density, and momentum density, respectively. The vector V^i will be called the *transport velocity*. Note that we have factored out

$$\hat{u}^t = \Gamma_2 \alpha^{-1}, \tag{34}$$

which has red-shift factors from both special (Γ_2) and general (α) relativity.

By substituting (33a) and (33c) in (17) we obtain our equation for mass density

$$\partial_t(D\sqrt{-g}) + \partial_i(DV^i\sqrt{-g}) = 0, \tag{35}$$

where $\sqrt{-g} = \alpha[\det(\gamma)]^{1/2}$ is the determinant of the four-metric. Note that the α in D and in $\sqrt{-g} \equiv \alpha[\sqrt{\det(\gamma)}]^{1/2}$ cancel. Thus, it is sometimes more convenient to evolve $\rho\Gamma_2$ than D (particularly if $\alpha \to 0$ in the strong field region). This equation is the familiar mass-continuity equation.

We rewrite (15) in the equivalent form

$$(\sqrt{-g})^{-1/2}\partial_b(T^b_a\sqrt{-g}) - \tfrac{1}{2}(\partial_a g_{cd})T^{cd} = 0, \tag{36}$$

and use our new variables (33) to write (9) as

$$T^a_b = S^a S_b/S^t + P\delta^a_b. \tag{37}$$

Taking the spatial component $a = i$ yields our momentum equation:

$$(\sqrt{-g})^{-1/2}[\partial_t(S_i\sqrt{-g}) + \partial_j(S_i V^j\sqrt{-g})] + \partial_i P - \tfrac{1}{2}(\partial_i g_{cd})S^c S^d/S^t = 0. \tag{38}$$

The last term gives the gravitational acceleration.

Finally, we use (17) in combination with $\hat{u}^a \nabla^b T_{ab} = 0$ to get our energy equation:

$$\partial_t(E\sqrt{-g}) + \partial_i(EV^i\sqrt{-g}) + P[\partial_t(\hat{u}^t\sqrt{-g}) + \partial_i(\hat{u}^t V^i\sqrt{-g})] = 0. \tag{39}$$

The last term is the change in internal energy caused by "*PdV*" work.

Note that by the choice of variables (33) our general relativistic hydrodynamic equations (35), (38), and (39) look very similar to nonrelativistic hydrodynamics. Special relativity is hidden in the Γ_2 terms in (D, E, S_i, $\hat{u}^t$). General relativity comes in through the space and time variation of $\sqrt{-g}$ and in the gravitational acceleration term in (38). Having evolved the fluid quantities, one can then reconstruct the source terms for Einstein's equations (12) simply by

$$d = \alpha^2 S^t - P, \tag{40a}$$

$$j_i = -\alpha S_i, \tag{40b}$$

$$S_{ij} = S_i S_j/S^t + P\gamma_{ij}. \tag{40c}$$

We then have a complete system of coupled geometrohydrodynamical equations in (6), (8), (35), (38), (39). (For their use in numerical calculations, see [17,43].)

B. The Comoving Limit

In general, β_i is chosen to simplify the gravitational field (see Smarr [42] and Wilson [43], for examples). However, we now show how choosing $\beta_i = \alpha v_i$ reduces our Eulerian hydrodynamics equations to the comoving ones discussed above. In the comoving limit (20), we have

$$D \rightarrow \rho(\alpha^2 - \beta^2)^{-1/2}, \tag{41a}$$

$$E \rightarrow \varepsilon\rho(\alpha^2 - \beta^2)^{-1/2}, \tag{41b}$$

$$V^i \rightarrow 0, \tag{41c}$$

$$S^t \rightarrow \rho h(\alpha^2 - \beta^2)^{-1}, \qquad S^i \rightarrow 0, \tag{41d}$$

$$S_i \rightarrow \rho h \beta_i (\alpha^2 - \beta^2)^{-1}. \tag{41e}$$

The primary simplification is that all transport terms (V^i terms) drop out because the spatial coordinates comove with the matter. Using (41a) and (41c) the continuity equation immediately reduces to (23). Besides losing the transport term in (38), the gravitational acceleration has only one term: $-\frac{1}{2}S^t\partial_i g_{tt}$. In the spherical case of comoving time slices with zero-shift vector, e.g., May and White, this term becomes, using (41d),

$$\tfrac{1}{2}S^t\partial_i g_{tt} \rightarrow \rho h\partial_i \ln \alpha = \rho h a_i, \tag{42}$$

where a_i is the spatial component of the four-acceleration vector of the (comoving) Eulerian observer. In this special case $S_i \rightarrow 0$, the only other term left in (38) is $\partial_i P$, yielding just the spatial components of the force equation (19). For this reason we refer to the last term in (38) as the gravitational acceleration term. In the general case, with $\beta_i \neq 0$, we can use comoving form of (35) and the first law (14), to show that our momentum equation becomes the comoving gauge equation (21). Finally, our energy equation reduces to (24) when one uses the mass continuity equation.

We point out that it is not necessary to use β_i to obtain a comoving system. Because one is using a finite-differenced grid in numerical calculations, one can move the grid points ($I, J = $ constant) relative to the spatial coordinate trajectories ($R, Z = $ constant). This can be represented as a *grid velocity* V_g^i, which can be set independent of β^i. In practice, V_g^i is used to have the grid move in a quasi-comoving sense. That is, it moves in as (nonspherical) collapse continues, so as to leave good spatial resolution of the flow, but it does not try to follow detailed swirling motion. The term V_g^i comes into the hydrodynamical equations by replacing V^i by $V^i - V_g^i$ in the transport terms. This method of computing hydrodynamics is often referred to as *mixed Euler–Lagrange*. (For details on its use, see Wilson [43].) If $V_g^i = V^i$, then the code becomes Lagrangian, while the shift vector β_i is free to set the spatial coordinates [52].

C.   *Einstein Equations*

Because we have *not* used up the shift vector freedom by imposing the comoving gauge, we can use it to simplify the gravitational field. This can be done by imposing conditions on the time derivatives of the three-metric functions. One way [2] minimizes the integrated coordinate shear on a time slice, whereas another [43] sets specific components of the three-metric to zero. The motivation in both cases is to separate out as much unnecessary coordinate shear from the true gravitational-wave degrees of freedom. For our discussion here, the important point is that such an approach leads to *elliptic* equations for β^i. Roughly speaking, this means that β^i will be determined by spatial *integrals* over the fluid variables, so that even if they suffer a step discontinuity, β^i will be continuous, unlike the comoving case.

Similarly, the use of maximal time slicing [39] has been shown to lead to smooth lapse functions α in a wide variety of cases [35,38,42–44]. Maximal slicing is obtained by demanding $\partial_t \, \text{tr}(K) = \text{tr}(K) = 0$, which when used in (7) yields an *elliptic* equation for α. Thus, this approach of freeing (α, β^i) from the detailed behavior of the matter flow corrects the first failure of comoving slices described in Section IV.B. Furthermore, as noted in that section, maximal time slicing usually avoids hitting spacetime singularities [2,35]. This allows one to study the subsequent accretion onto black holes after their formation. As mentioned, Chrzanowski (38) showed one could combine maximal slicing (α) with comoving shift (β^i) to get rid of the second failure of the comoving gauge.

This use of elliptic equations to solve for *nondynamical* or longitudinal parts of the spacetime metric can be extended to the three-metric and extrinsic curvature. Thus, in the isothermal gauge [42,43], one has only two nonzero components of γ_{ij} and three nonzero components of K_{ij} in a nonrotating axisymmetric spacetime. The Hamiltonian constraint (8b) can be converted into an elliptic equation for one of the γ_{ij}s (Wilson [43], Eq. 47) and the momentum constraint (8a) can be used to obtain two of the K_{ij}s by elliptic equations (Wilson [43], Eqs. 28 and 49). This leaves one γ_{ij} and one K_{ij}, which satisfy a wave equation (Wilson [43], Eqs. 50 and 51).

D.   *A Numerical Example of Nonspherical Collapse*

We turn finally to the third problem of comoving coordinates that occurs only for nonspherical flows. This problem of tangling of coordinate lines arises because: "After passing through a shock wave, potential flow [isentropic, irrotational flow] of a gas usually becomes rotational flow." This statement is from Landau and Lifshitz's [12] first page on two-dimensional gas flow. We quote it only to emphasize how ubiquitous this phenomenon is.

To illustrate this effect, we present here some results on the gravitational collapse and bound of a highly distorted star. This calculation was done using the mixed Euler–Lagrange scheme just described and is discussed by Wilson [43]. The equation of state is given by

$$P = \rho \varepsilon \tag{43}$$

appropriate for stiff matter ($\gamma = 2$), such as a neutron star near bounce. The initial condition was the star was axisymmetric and nonrotating with density profile

$$\rho = \rho_0(\sin kx/x), \qquad x = (\sqrt{R^2 + 2Z^2})^{1/2}, \tag{44}$$

appropriate for a $n = 1$ polytrope. The internal energy was 10% of that required to support the star in the Z direction. (We use cylindrical coordinates Z, R, ϕ.)

The star collapsed from rest and bounced at $T \sim 53M$ ($G = C = 1$) when the central α had fallen to $\alpha_c \sim 0.45$ (see Fig. 7 in Wilson [43]). This was therefore a very relativistic bounce. We present here the variation of the density D and the velocity field V^i as functions of (R, Z, T). Figure 1 shows the contours of density and the velocity field at $T \sim 20M$. One can see that the density contours are flattened oblate spheroids and the matter flow is essentially radial. As the center slows down to bounce, the outer layers run into the core forming a strong shock ($T \sim 45M$), which propagates outward. By $T \sim 70M$, Fig. 2, the shock is rapidly accelerating in the decreasing density gradient. Note that the density contours show that the star has bounced along the Z axis and the core is now highly prolate. Most important one sees that the matter flow behind the shock has become rotational with counterrotating vortex rings above and below the equatorial plane ($Z = 0$). Recall the total angular momentum must still be zero since there is no rotation around the Z axis. By $T \sim 130M$, Fig. 3, the shock has reversed almost the entire velocity field. The vortex ring has moved outward, and in its wake has left another vortex ring rotating in the opposite direction. As the prolate core stops and begins to fall inward along the Z axis and outward along the equator, this vortex is reinforced and remains present until the end of the calculation ($T \sim 250M$).

If this calculation has been done with comoving coordinates, the spatial coordinate trajectories would have ended up in a number of swirling vortex rings. As they stretch and twist the metric functions γ_{ij} measuring the proper distance between the coordinate lines would have developed wild distortions, *even though the gravitational field is fairly smooth*. This would severely complicate the extraction of gravitational radiation information as well as

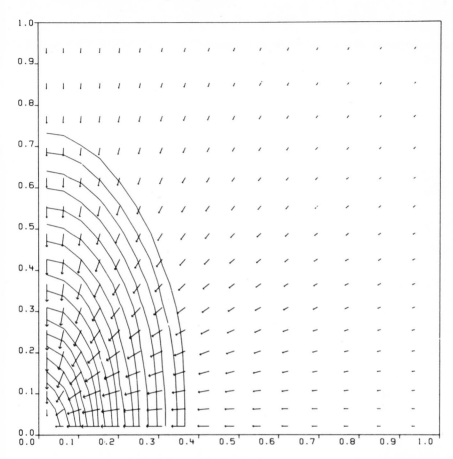

Fig. 1 The collapse of a distorted axisymmetric nonrotating star with a stiff equation of state. The Z axis of symmetry is to the right, while the equatorial plane is vertical. The distances and time scales are in units of $M (G = c = 1)$. This is a snapshot at $T \simeq 20M$ after the collapse from rest started. The density D contours show the initial 2:1 oblate distortion. The arrows represent the velocity vectors V^i of the fluid which at this time is in quasiradial infall (cycle $= 50$).

distorting its propagation through this grid. By contrast, in the mixed Euler–Lagrange approach, the grid velocity is used to ensure a uniform coverage of the active region. One can see this explicitly since the tails of the velocity vectors are at the grid points. At the same time the shift vector, which would be discontinuous at the shock front in the comoving gauge, is smoothly keeping $\gamma_{RZ} = 0 = \gamma_{RR} - \gamma_{ZZ}$ so that only two metric functions ($\gamma_{RR} = \gamma_{ZZ}$, $\gamma_{\phi\phi}$) need be evolved (instead of the four required in comoving gauge for this problem).

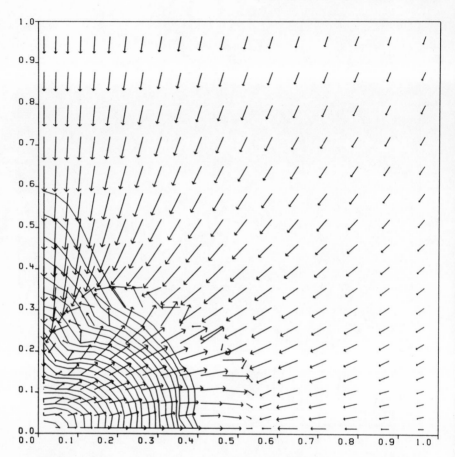

Fig. 2 The same collapse as Fig. 1 now at $T \sim 70M$. The core has bounced in the Z direction and is moving outward causing a shock wave. Behind the shock a swirl has developed due to the entropy gradients produced. The core is now prolate (cycle = 250).

E.   Vortices and Supernova Explosion

We shall turn in a moment to a quantitative description of the growth of circulation following a shock, but first we remark on the possible astrophysical significance of such model calculations. Using the most up-to-date physics, one can calculate very detailed *spherical* models of core collapse (see, e.g., Wilson [43]). These include accurate hot equations of state, detailed neutrino physics, and transport. However, the result is that it is not clear whether the shock that forms on bounce can drive off the envelope with certainty. Therefore, it is crucial to determine if deviations from sphericity could introduce new effects.

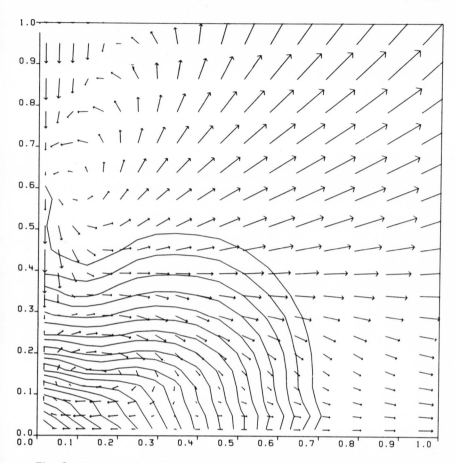

Fig. 3 The collapse is at T ~ 130M. The accelerating shock wave has swept off the grid leaving two counterrotating vortices behind it. The core is just about to bounce on the equator (cycle = 450).

Three recently discussed mechanisms may be active in producing a nonspherical configuration after the first bounce in stellar core collapse. First, Saenz and Shapiro [45] have shown that at least under very simple conditions the eccentricity of a bouncing stellar core grows rapidly with each bounce. Second, Epstein [6] has pointed out that after a stellar core has formed, the outside neutronizes rapidly and becomes convectively unstable due to high lepton number on the inside and low lepton number on the outside. If this process overshoots, Colgate [7] has argued that a large-scale Rayleigh–Taylor instability will take place, completely turning over the core by the third bounce. Third, these effects may be enhanced by a peculiar condition of the core as calculated in the most recent spherical runs [51]. It is found that when matter at densities above $\sim 10^{13}\, \text{g cm}^3$ does neutronize, the pressure in the matter may *increase* if the concomitant neutrino losses are not too high. After bounce at $\sim 10^{14}\, \text{g/cm}^3$, the core density falls toward $\sim 10^{13}\, \text{g/cm}^3$. If convection exists and enables a piece of the core to lose some neutrinos, its pressure will rise and it will acquire further bouyancy. Thus, if these three mechanisms can work in concert, it might be possible to reduce large vortical motion in the core.

One of us (JRW) is using a detailed spherical supernova code to see if these particular convection processes work with the restriction to near-spherical symmetry. However, we want to point out here that even in a purely hydro-dynamical collapse (no neutrino transport) with the simplest (unrealistic) equation of state, *the nonspherical shock itself drives convection strongly*. Note that two swirls have already occurred (Fig. 3) before the second bounce. It seems likely that this behavior will occur in any nonspherical collapse. Note that in purely spherical collapse the shock front is at right angles to the flow and this induces no circulation. This generic effect may also help Shapiro's mechanism [45] for amplification of core eccentricities during subsequent core bounces. We are currently studying this problem in more detail.

F.   The Circulation Theorem

In order to quantify this generation of vortices we briefly review Taub's [5] relativistic generalization of the classical circulation theorem. (For the nonrelativistic case see, e.g., Landau and Lifshitz [12].) Consider a closed contour with path parameter τ and spacelike tangent vector λ^a. Then the circulation around that contour is

$$C = \oint h\hat{u}_a \lambda^a\, d\tau. \tag{45}$$

Note that the integrated quantity is the enthalpy weighted four-velocity. Let the contour be carried into the future along the fluid lines. Then if s is the proper time along the fluid lines, the time rate of change of C is

$$\frac{dC}{ds} = \oint TS_{,a}\lambda^a\, d\tau, \tag{46}$$

that is, entropy gradients can drive the growth of circulation.

Taub defines a convenient antisymmetric tensor:

$$\Omega_{ab} = \nabla_a(h\hat{u}_b) - \nabla_b(h\hat{u}_a) \tag{47a}$$

$$= -2h\omega^d\hat{u}^c\eta_{dcab} + T(\hat{u}_b S_{,a} - \hat{u}_a S_{,b}), \tag{47b}$$

where η_{dcab} is the alternating tensor and ω^d is the vorticity vector:

$$\omega^d = \tfrac{1}{2}\eta^{dcab}\hat{u}_c\,\hat{u}_{a,b}. \tag{48}$$

By Stokes theorem, C is equivalent to the area integral over Ω_{ab}. Thus, the initial circulation vanishes only if the flow is *both* irrotational and isentropic, i.e., potential flow. Applying Stokes theorem to (46) one can show that the circulation is conserved only if the fluid is barotropic [5].

As Taub [5] describes, this cannot be the case when shocks are present. Even if the flow is isentropic before the shock occurs, a general shock front will leave an entropy gradient behind it, which will drive circulation, just as we saw in Figs. 1–3. Furthermore, if the hydrodynamics is nonspherical, the microphysics will evolve nonspherically, setting up additional entropy gradients. We are studying our collapses to verify that they quantitatively agree with (46).

We note in passing that all of this analysis can be performed very elegantly using differential forms (see Carter [46]). He finds that the quantity $hu_a \equiv \pi_a$ is the natural momentum for a fluid. His circulation theorem (3.9) states that the circulation is *always* conserved *if* one uses as time derivative not proper time $\hat{u}^a\nabla_a$, but rather $1/T\hat{u}^a\nabla_a$.

VI. The Velocity Potential Formalism

Let us summarize the methods for solving hydrodynamics. In the co-moving approach (Section IV) the three spatial components of the fluid's four-velocity are set equal to zero. The fourth component is fixed by $\hat{u}^a\hat{u}_a = -1$. The entropy S is constant along each world line and the rest-mass density ρ is obtained from the variations in the spacetime metric from (23). The energy density ε is calculated from the first law (13, 24) with $P = P(\rho, S)$ given by the equation of state. In the Eulerian approach ρ and ε are evolved as D and E in Eqs. (35) and (39). The entropy is then calculated from the first law with $P = P(\rho, \varepsilon)$ as the equation of state. The spatial components of the four-velocity are solved for as S_i the momentum density, with the fourth component again set by the four-velocity normalization.

The velocity potential formalism is a third method, which combines some features of both comoving and Eulerian methods. In this approach one regards the fluid as represented by a set of six scalars on spacetime. These scalars satisfy evolution equations similar to the Eulerian ones. However, the scalars are chosen in such a way that most of them are constant along the fluid worldlines. The formalism goes back to Clebsch [47] as updated by Lin [48] and Seliger and Whitham [49]. Schutz [8] worked out the general relativistic version that we rewrite in our language. See also [53].

The six scalars will be denoted by $(\rho, S, \theta, \phi, \chi, \xi)$. The first two are just the rest-mass density and entropy. We shall assume all other thermodynamic quantities (ε, P, h, T) are functions of (ρ, S) and can be obtained by equation of state or the first law. The other four scalars represent the four components of the four-velocity in the following representation:

$$\hat{u}_a = h^{-1}[\partial_a \phi + \theta \partial_a S + \chi \partial_a \xi]. \tag{49}$$

In the case of isentropic irrotational flow, this reduces to the standard representation of potential flow $V = \nabla \phi$ discussed, e.g., in Landau and Lifshitz [12]. Allowance for entropy gradients is made with (θ, S) and for vorticity by (χ, ξ).

The evolution equations for the scalars (see Eq. (2.22) in Schutz [8]), are rewritten in our Eulerian form by use of (1c) to write the fluid proper time derivative as

$$\hat{u}^a \partial_a = \Gamma_2 \alpha^{-1}(\partial_t + V^i \partial_i). \tag{50}$$

The six equations become

$$\partial_t(D\sqrt{-g}) + \partial_i(DV^i\sqrt{-g}) = 0, \tag{51a}$$

$$\partial_t S + V^i \partial_i S = 0, \tag{51b}$$

$$\partial_t \chi + V^i \partial_i \chi = 0, \tag{51c}$$

$$\partial_t \xi + V^i \partial_i \xi = 0, \tag{51d}$$

$$\partial_t \phi + V^i \partial_i \phi + \alpha \Gamma_2^{-1} h = 0, \tag{51e}$$

$$\partial_t \theta + V^i \partial_i \theta - \alpha \Gamma_2^{-1} T = 0. \tag{51f}$$

To evolve from one time slice to next, one takes the scalars and their spatial derivatives and constructs the spatial components of $\hat{u}_i$. From these and the metric $(\alpha, \beta^i, \gamma_{ij})$, one easily calculates

$$\Gamma_2 = (1 + \hat{u}_i \gamma^{ij} \hat{u}_j)^{1/2}, \tag{52a}$$

$$v^i = \hat{u}_i \Gamma_2^{-1}, \tag{52b}$$

$$V^i = \alpha v^i - \beta^i. \tag{52c}$$

These then allow the calculation of the time derivatives for all six scalars in (51).

In the comoving limit we recover (22) and (23) for S and ρ, plus

$$\chi = \chi(x^i), \tag{53a}$$

$$\xi = \xi(x^i), \tag{53b}$$

$$\partial_t \phi + (\alpha^2 - \beta_k \beta^k)^{1/2} h = 0, \tag{53c}$$

$$\partial_t \theta - (\alpha^2 - \beta_k \beta^k)^{1/2} T = 0, \tag{53d}$$

such that again the hydrodynamics is automatically integrated. The proof that the velocity potential evolution equations are equivalent to the standard ones is given in Schutz [8].

The circulation theorem can be rewritten in this language (see Schutz [8]). The circulation itself is given by

$$C(s) = \oint [\alpha \beta_{,a} \lambda^a + \theta S_{,a} \lambda^a] d\tau,$$

which clearly shows that it is composed of vorticity (α, β) and thermal circulation (θ, S). The time derivative (46) is easily obtained since $dC/ds = \hat{u}^a \partial_a C$ and $\hat{u}^a \partial_a$ on (α, β, S) vanish, while $\hat{u}^a \partial_a \theta = T$.

The velocity potential representation has not been used much for numerical calculations. It is not clear whether it would be more useful than the Eulerian approach. The major area that should be worked out before it can be implemented is how to handle shocks. Both the comoving [28] and the Eulerian [43] schemes use artificial viscosity. The justification for this is given by May and White [28] by first working out the equations for an *imperfect* fluid, which allows for viscosity and heat transport. Such a calculation would be necessary for the velocity potential scheme (presumably adding new scalars), before one would be sure how to add artificial viscosity. We are currently investigating this possibility.

VII. Conclusions

We have attempted to present a unified treatment of the methods researchers have used to study relativistic hydrodynamics. Our emphasis has been on the form of the equations and the kinematic implications of gauge choices because these have important consequences for numerical calculations. The interplay between theoretical formalisms and "numerical experiments" is nicely illustrated by the growth of circulation behind shocks. These results may have interesting astrophysical applications that we shall turn to in a future paper.

Acknowledgments

We particularly wish to thank Abe Taub for his relentless questions and stimulating discussions over the last few years, which directly led to this paper. We have benefited from conversations with B. Carter, D. M. Eardley, K. R. Eppley, J. LeBlanc, V. Moncrief, S. A. Colgate, and J. W. York, Jr. One of us (LS) wishes to thank the Institute of Astronomy, Cambridge, England, for its hospitality during a portion of this research. Some of this work was performed under the auspices of the U.S. Department of Energy under contract W-7405-ENG-48. Partial support was also given by the National Science Foundation under grant numbers PHY77-18762 and PHY78-09616. An earlier stage of this work was done at Princeton University Observatory.

References

[1] Arnowitt, R., Deser, S., and Misner, C. W., in "Gravitation: An Introduction to Current Research" (L. Witten, ed.). Wiley, New York, 1962.

[2] Smarr, L., and York, J. W., Jr. Phys. Rev. D 17, 2529 (1978); York, J. W., Jr., in "Sources of Gravitational Radiation" (L. L. Smarr, ed.). Cambridge Univ. Press, London and New York, 1979.

[3] May, M. M., and White, R. H., Phys. Rev. 141, 1232 (1966).

[4] Pachner, J., Bull. Astron. Inst. Czech. 19, 33 (1968); J. Comp. Phys. 15, 385 (1974); in "General Relativity and Gravitation" (G. Shaviv and J. Rosen, eds.). Wiley, New York, 1979.

[5] Taub, A. H., Arch. Ration. Mech. Anal. 3, 312 (1959).

[6] Epstein, R., Mon. Not. R. Astron. Soc. (in press).

[7] Colgate, S. A., "Supernova Mass Ejection and Core Hydrodynamics." Prepr. LA-UR-78-2030. Los Alamos Sci. Lab., Los Alamos, New Mexico, 1978.

[8] Schutz, B. F., Jr., Phys. Rev. D 2, 2762 (1970).

[9] Smarr, L., Wilson, J. R., and Piran, T., "Dragging Regions in Relativistic Space-times, preprint, 1980.

[10] Smarr, L. L., Ph.D. Thesis, Univ. of Texas, Austin, 1975 (and references therein).

[11] Thorne, K. S., in "High Energy Astrophysics" (C. DeWitt, E. Saltzman, and P. Vernon, eds.), Vol. 3. Gordon & Breach, New York, 1967.

[12] Landau, L. D., and Lifshitz, E. M., "Fluid Mechanics." Pergamon, Oxford, 1969.

[13] Taub, A. H., Annu. Rev. Fluid Mech. 10, 301 (1978).

[14] Taub, A. H., Lect. Appl. Math. 8, 170 (1970).

[15] Taub, A. H., Phys. Rev. 74, 328 (1948).

[16] Thorne, K. S., Astrophys. J. 179, 897 (1973).

[17] Wilson, J. R., Astrophys. J. 173, 431 (1972); Astrophys. J. 176, 195 (1972); N.Y. Acad. Sci. 262, 123 (1975); Ruffini, R., and Wilson, J. R., Phys. Rev. D 12, 2959 (1975); Wilson, J. R., in "Proceedings of the First Marcel Grossman Meeting on General Relativity" (R. Ruffini, ed.). North-Holland Publ., Amsterdam, 1977; Proc. Int. Sch. Phys. "Enrico Fermi" 65 (1978).

[18] Tauber, G. E., and Weinberg, J. W., Phys. Rev. 122, 1342 (1961).

[19] Tolman, R. C., "Relativity, Thermodynamics, and Cosmology." Oxford Univ. Press (Clarendon), London, 1934.

[20] Taub, A. H., Ill. J. Math. 1, 270 (1957).

[21] MacCallum, M. A. H., and Taub, A. H., Commun. Math. Phys. 25, 173 (1972).

[22] Tolman, R. C., Proc. Natl. Acad. Sci. U.S.A. 20, 164 (1934).

[23] Bondi, H., Mon. Not. R. Astron. Soc. 107, 410 (1947).

[24] Misner, C. W., and Sharp, D. H., Phys. Rev. 136, B571 (1964).

[25] Misner, C. W., and Sharp, D. H., Phys. Lett. 15, 279 (1965); Misner, C. W., Phys. Rev. 137, B1360 (1965); in "Astrophysics and General Relativity" (M. Chretien, S. Deser, and J. Goldstein, eds.), Vol. 1. Gordon & Breach, New York, 1969.

[26] Bardeen, J. M., Ph.D. Thesis, Calif. Inst. Technol. Pasadena, California, 1965.

[27] Podurets, M. A., Sov. Astron.—AJ 8, 19 (1964); Sov. Phys.—Dokl. 9, 1 (1964).

[28] May, M. M., and White, R. H., Methods Comput. Phys. 7, 219 (1967).

[29] Schwarz, R. A., Ann. Phys. (N.Y.) 43, 42 (1967).

[30] Wilson, J. R., Astrophys. J. 163, 209 (1971).

[31] Matsuda, T., and Nariai, H., Prog. Theor. Phys. 49, 1195 (1973).

[32] Voropinov, A. I., Zaguskin, V. L., and Podurets, M. A., Sov. Astron.—AJ, 11, 442 (1967).

[33] Matsuda, T., and Sato, H., Prog. Theor. Phys. 41, 1021 (1969).

[34] Appenzeller, I., and Fricke, K., *Astron. Astrophys.* **18**, 10 (1972).

[35] Eardley, D. M., and Smarr, L., *Phys. Rev. D* (1979).

[36] Marsden, J. E., and Tipler, F. J., "Maximal Hypersurfaces and Foliations of Constant Mean Curvature in General Relativity," Prepr. Univ. of California, Berkeley, 1978.

[37] Wheeler, J. A., *in* "Relativity, Groups, and Topology" (C. DeWitt and B. S. DeWitt, eds.). Gordon & Breach, New York, 1964.

[38] Chrzanowski, P., *Yale Workshop Dyn. Constr. Spacetime, 1977*; Chrzanowski, P., and Piran, T., preprint, 1979.

[39] Lichnerowicz, A., *J. Math. Pure Appl.* **23**, 37 (1944); Estabrook, F., Wahlquist, H. D., Christensen, S., DeWitt, B., Smarr, L., and Tsiang, E., *Phys. Rev. D* **7**, 2814 (1973).

[40] Novikov, I. D., and Polnarev, A. G., *in* "Sources of Gravitational Radiation" (L. L. Smarr, ed.). Cambridge Univ. Press, London and New York, 1979.

[41] Lindquist, R. W., *Ann. Phys. (N.Y.)* **37**, 487 (1966).

[42] Smarr, L., *in* "Sources of Gravitational Radiation" (L. L. Smarr, ed.). Cambridge Univ. Press, London and New York, 1979.

[43] Wilson, J. R., *in* "Sources of Gravitational Radiation" (L. L. Smarr, ed.). Cambridge Univ. Press, London and New York, 1979.

[44] Eppley, K. R., Ph.D. Thesis, Princeton Univ., Princeton, New Jersey, 1975; Smarr, L., *Ann. N.Y. Acad. Sci.* **302**, 569 (1977); Duncan, M., "Maximally Slicing a Charged Black Hole with Minimal Shear," Prepr. Univ. of Texas, Austin, 1977; Piran, T., *Phys. Rev. Lett.* **41**, 1085 (1978); Piran, T., *in* "Sources of Gravitational Radiation" (L. L. Smarr, ed.). Cambridge Univ. Press, London and New York, 1979.

[45] Shapiro, S. L., *in* "Sources of Gravitational Radiation" (L. L. Smarr, ed.). Cambridge Univ. Press, London and New York, 1979.

[46] Carter, B., "Perfect Fluid and Magnetic Field Conservation Laws in the Theory of Black Hole Accretion Rings." Preprint, 1978.

[47] Clebsch, A., *J. Reine Angew. Math* **56**, 1 (1859).

[48] Lin C. C., *Proc. Sch. Phys. "Enrico Fermi"* **21** (1963).

[49] Seliger, R. L., and Whitham G. B., *Proc. R. Soc. London Ser. A* **305**, 1 (1968).

[50] Lichnerowicz, A., "Relativistic Hydrodynamics and Magnetohydrodynamics." Benjamin New York, 1967.

[51] Wilson, J. R., *Ann. N.Y. Acad. Sci.* **336**, 358 (1980).

[52] T. Piran, *J. Comp. Phys.* (in press).

[53] Tam, K., and O'Hanlon, J., *Nuovo Cimento* **62B**, 351 (1966); Ray, J. R., *J. Math. Phys.* **13**, 1451 (1972).

12

Lagrangian Relativistic Hydrodynamics with Eulerian Coordinates

Tsvi Piran†

Center for Relativity
University of Texas at Austin
Austin, Texas

Abstract

I describe a new formulation of Lagrangian general relativistic hydrodynamics. The approach is based on Eulerian coordinates and a family of Lagrangian observers. This formulation provides the simplicity of Lagrangian hydrodynamics but avoids the drawbacks of Lagrangian coordinate systems. It is described with numerical applications in mind.

I. Introduction

The Lagrangian formulation of hydrodynamics provides a simplicity not found in the rival Eulerian approach. Taub was the leader in extending the Lagrangian formalism to general relativistic hydrodynamics (see, e.g., [1]). In this paper I shall be concerned with the further development of his approach, and I shall attempt to overcome difficulties that have been found in the current formalism.

The relationship between coordinates and geometry in general relativity poses problems for relativistic Lagrangian hydrodynamics that do not exist in the Newtonian formulation. Smarr *et al.* (Chap. 11, this volume) find that the Lagrangian condition imposes severe limitations on the coordinates. The

---

† Present address: Institute for Advanced Study, Princeton, New Jersey 08540.

Lagrangian coordinates become inadequate for describing the underlying geometry. This, in turn, limits the potential usage of the relativistic Lagrangian approach. Similar problems do not arise using an Eulerian formulation of relativistic hydrodynamics.

I describe here an attempt to reformulate Lagrangian relativistic hydrodynamics that avoids the drawbacks of the current formalism. This approach combines Lagrangian hydrodynamics with Eulerian† coordinates. It stems from the "grid velocity" concept, introduced to numerical general relativity by Wilson [2,5] (see also Smarr et al., Chap. 11, this volume). The Eulerian coordinates are free of any hydrodynamic constraints. A family of comoving observers recovers the Lagrangian nature of the hydrodynamic equations. The observers' motion causes the flux terms to disappear from the hydro-dynamic equations. These observers, but not the coordinates, follow the matter. All geometrical and hydrodynamic quantities are measured by the comoving observers relative to the local tetrads defined by the Eulerian coordinate system. The resulting scheme retains the original Lagrangian simplicity without the drawbacks of the Lagrangian coordinates.

I use the ADM [3] formulation of general relativity and the notations of Smarr et al (Chap. 11, this volume): $i, j = 1, 3; a, b = 1, 4$ with 4 the timelike component. The geometry is described by the three metric g_{ij}, the extrinsic curvature K^i_j, the lapse function α and the shift vector β^i. For illustrative purposes, the discussion is limited to perfect fluids, i.e.,

$$T^a_b = [\rho(1 + e) + p]u^a u_b + p\delta^a_b, \tag{1}$$

where T^a_b is the energy momentum tensor, u^a the four velocity of the fluid, ρ, e, and p are the rest-mass density, specific-energy density, and pressure, respectively, and $S_i = [\rho(1 + e) + p]u_i u^t$ is the momentum density of the fluid. Geometric coordinates $c = 8\pi G = 1$ are used.

II. General Formalism

I define general Eulerian coordinates (x^i, t) on space–time and a family of Lagrangian observers, labeled by y^i, whose coordinates are $x^i(y^i, t)$. The motion of the Lagrangian observers is defined by

$$\partial x^i(y^i, t)/\partial t \equiv V^i_L(y^i, t). \tag{2}$$

† I define Lagrangian relativistic coordinates (observers) as comoving with the matter field. Relativistic Eulerian coordinates (observers) are any non-Lagrangian coordinates (observers). This definition of relativistic Eulerian coordinates differs from the definition given by Smarr et al. (Chap. 11, this volume).

A general function, vector, or tensor component $f(x^i, t)$ can be expressed as a function of the Lagrangian observers:

$$F(y^i, t) \equiv f(x^i(y^i, t), t). \tag{3}$$

These observers measure all quantities with respect to the local Eulerian tetrads, therefore the observers' motion results only in an addition of a Lagrangian velocity term $V_L^i(\partial/\partial x^i)f$ to the time derivative of F:

$$\frac{\partial F(y^i, t)}{\partial t} = \frac{\partial f}{\partial t} + V_L^i \frac{\partial}{\partial x^i} f. \tag{4}$$

This should be compared with a Lie derivative term $\mathscr{L}_\beta F$ that results from the coordinate motion induced by the shift vector.

A. Geometry

The modified ADM equations for the geometry are

$$\frac{\partial}{\partial t} g_{ij}(y^i, t) = -2\alpha K_{ij} + \beta^k g_{ij,k} + g_{ik}\beta^k_{,j}$$
$$+ g_{jk}\beta^k_{,i} + V_L^k g_{ij,k}, \tag{5}$$

$$\frac{\partial}{\partial t} K_i^j(y^i, t) = -\alpha_{|i}^j + \alpha[KK_i^j + R_i^j + g^{kj}T_{ik}]$$
$$+ \beta^k K_{i,k}^j + K_k^j \beta^k_{,i} - K_i^k \beta^j_{,k} + V_L^k K_{i,k}^j, \tag{6}$$

where | denotes a covariant derivative relative to the three-metric g_{ij}.

B. Hydrodynamics

Following Wilson [4] a general form of the Eulerian hydrodynamic equation is

$$(\partial Q/\partial t) + (\partial/\partial x^i)(QV^i) = G_Q, \tag{7}$$

where $V^i \equiv u^i/u^t$ and Q stand for the generalized particle number $D \equiv (\rho u^t \sqrt{-g})$; $[\sqrt{-g}$ is the determinant of the four-metric, i.e., $\sqrt{-g} \equiv \alpha(\sqrt{(3)}g)^{1/2}]$, generalized energy density $E \equiv (eu^t \sqrt{-g})$, and generalized components of the momentum density vector $S_i \equiv (\rho(1 + e) + p)u_i u^t \sqrt{-g}$. The addition of the Lagrangian velocity term to this equation yields

$$\frac{\partial Q(y^i, t)}{\partial t} + \frac{\partial}{\partial x^i} [(V^i - V_L^i)Q] + Q \frac{\partial}{\partial x^i} V_L^i = G_Q(y^i, t). \tag{8}$$

The Lagrangian velocity is now defined so that the flux term $(\partial/\partial x^i) \times [(V^i - V^i_L)Q]$ vanishes:

$$V^i_L \equiv V^i = \frac{g^{ij}S_j}{[\rho(1 + e) + p]u^{t^2}} - \beta^i.$$

(9)

Using

$$\frac{\partial}{\partial x^i} V^i_L = \frac{\partial}{\partial t} \ln\left(\frac{\partial x^i}{\partial y^i}\right),$$

(10)

the hydrodynamic equations are cast into the Lagrangian form:

$$\frac{\partial}{\partial t}\left[Q(y^i, t)\left(\frac{\partial x^i}{\partial y^i}\right)\right] = \left(\frac{\partial x^i}{\partial y^i}\right)G_Q(y^i, t).$$

(11)

Note that $\tilde{Q} \equiv Q(\partial x^i/\partial y^i)$ is the corresponding Q measured in the local frame of the Lagrangian observers, i.e., with respect to the y^i coordinates.

The final forms of the hydrodynamic equations are

$$\frac{\partial}{\partial t}\left(D\frac{\partial x^i}{\partial y^i}\right) = 0,$$

(12)

$$\frac{\partial}{\partial t}\left(E\frac{\partial x^i}{\partial y^i}\right) + p\frac{\partial}{\partial t}\left(\sqrt{-g}u^t\frac{\partial x^i}{\partial y^i}\right) = 0,$$

(13)

$$\frac{\partial}{\partial t}\left(S_j\frac{\partial x^i}{\partial y^i}\right) + \left(\frac{\partial x^i}{\partial y^i}\right)\left[\sqrt{-g}\frac{\partial p}{\partial x^j} - \frac{S^t}{2}\frac{\partial\alpha^2}{\partial x^j} + S_i\frac{\partial}{\partial x^j}\beta^i - \frac{S_iS_k}{2S^t}\frac{\partial}{\partial x^j}g^{ik}\right] = 0.$$

(14)

V^i and the flux term in Eq. (8) vanish for a suitable choice of the shift vector [Eq. (9)]. This is the traditional Lagrangian coordinate condition. The resulting equations are equivalent to the equation for $\tilde{Q}$. In other words, the term $Q(\partial/\partial x^i)V^i_L$, which appears in this formalism but does not appear in the traditional Lagrangian formulas, expresses the fact that Q is not observed with respect to the Lagrangian coordinates. The different relation between the effects of Lagrangian velocity and the shift vector on the geometric and hydrodynamic equations stems from the fact that the geometric equations were projected along the normals to the constant time slices, while the hydrodynamic equations are projected along u^a.

III. Numerical Scheme

Equations (5)–(6), (12)–(14) are supplemented by an equation of state and by arbitrary conditions for the lapse function and the shift vector to form a complete set of equations.

The Lagrangian coordinates y^i become the numerical grid points of the finite differencing scheme. Thus we replace y^i by $I^i = (J, K, L)$, which takes only integer values. The grid points comove with the matter so that the numerical grid cells correspond to Lagrangian fluid elements. It is clear now that the Lagrangian velocity V_L^i is a special case of Wilson's [2] grid velocity V_g^i. The coordinate transformation factor $(\partial x^i / \partial y^i)$ corresponds to the grid spacing, i.e.,

$$\left(\frac{\partial x^i}{\partial y^i}\right) \to \prod_{i=1}^{3} \delta X^i(I^i) = \prod_{i=1}^{3} (X_{I^i + 1/2}^i - X_{I^i - 1/2}^i), \qquad (15)$$

where δ is the finite differencing centered derivative operator. In practice [5] Eq. (11) is solved every time step for $\tilde{Q}$ and it is rescaled subsequently by

$$\prod_{i=1}^{3} (X_{I^i + 1/2}^i - X_{I^i - 1/2}^i)_{\text{old}} / (X_{I^i + 1/2}^i - X_{I^i - 1/2}^i)_{\text{new}}. \qquad (16)$$

The Lagrangian numerical grid must be supplemented with an exterior vacuum grid with a vanishing Lagrangian velocity, thus allowing for evaluation of the geometry in this region. The matching procedure is relatively simple since the same Eulerian coordinates are defined on both grids. A second anticipated problem is accumulation of the Lagrangian grid points behind shock fronts. This problem can be solved by a devision and rezoning procedure.

Acknowledgments

I am grateful to J. Wilson for many helpful discussions. This research was supported by NSF grant PHY77-22489.

References

[1] Taub, A. H., *Annu. Rev. Fluid Mech.* **10**, 301 (1978).
[2] Wilson, J. R., Lecture, GR8, Waterloo, Canada, 1977 (unpublished).
[3] Arnowitt, R., Deser, S., and Misner, C. W., *in* "Gravitation: An Introduction to Current Research" (L. Witten, ed.), p. 227. Wiley, New York, 1962.
[4] Wilson, J. R., *Astrophys. J.* **173**, 431 (1972).
[5] Wilson, J. R., *in* "Sources of Gravitational Radiation" (L. Smarr, ed.), p. 423. Cambridge Univ. Press, London and New York, 1979.

13

Some Thoughts on the Origin of Cosmic Inhomogeneities†

E. P. T. Liang

Institute of Theoretical Physics
Department of Physics
Stanford University
Stanford, California

Abstract

The fundamental problems associated with a physical origin of cosmic inhomogeneities are discussed. Two alternative pictures are considered: the selective survival of random chaotic fluctuations by nonlinear hydrodynamic effects and a possible gravithermal first-order transition at decoupling. Current results do not favour a completely chaotic hydrodynamic picture.

I. Introduction

In this paper I try to reconsider some of the fundamental difficulties associated with formulating a physical theory of the origin of cosmic inhomogeneities and suggest several possible new approaches. It is not intended to be a comprehensive review since several excellent ones are available on the subject [1–5]. Rather, it will be a more systematic organization of some scattered ideas that I have recently privately advocated but not publicized.

Simply stated, the dilemma of cosmic inhomogeneities is twofold.

(a) If the Universe was indeed initially smooth as indicated by the isotropy of the microwave background, the abundance of the light elements,

† Work partially supported by NSF Grant PHY 76–21454.

191

etc., it appears very difficult, if not impossible, for macroscopic hydrodynamic fluctuations to arise spontaneously in the fireball (predecoupling, redshift $z > 1400$) phase and grow into finite amplitude inhomogeneities by decoupling because of causality constraints (existence of particle horizons) and suppression of gravity by radiation pressure.

(b) If the Universe was initially chaotic with large random density and velocity perturbations on all scales, then it is hard to see why the fluctuation level today has a hierarchical structure with density contrast $\delta\rho/\rho$ rapidly decreasing with increasing scale instead of the opposite, since both dissipation and radiation pressure strongly favor the survival and growth of large clumps over the small ones.

That the initial conditions could be set "just right" to explain what we see is, at best, a metaphysical conjecture to be abhorred by serious physicists. Unlike the problem of the origin of the large specific entropy of the universe $S_b \sim 10^9$ (or, equivalently, the origin of the small baryon asymmetry), which may find its ultimate solution in the unknown physics of the very early Universe, the problem of cosmic inhomogeneities apparently must be answered in classical, ordinary terms. The times when macroscopic inhomogeneities enter the particle horizon are so late (e.g., $t_{\text{horizon} = \text{galaxy}} \sim 1$ year) that the physics must be conventional.

Section II summarizes the key observations with bearing on the theoretical problem. In Section III, I discuss the several critical mass scales relevant to the evolution of linear density perturbations in a Friedmann universe. Section IV deals with nonlinear evolutions of inhomogeneities in the fireball phase. Finally, I discuss the possible gravithermal origin of the inhomogeneities in Section V. For the readers not already familiar with cosmology, basic background material can be found in, for example, the books by Peebles [6], Weinberg [7] or the introductory article by Liang and Sachs [8].

II. Current Observational Results

All observational evidences indicate that the matter distribution is hierarchically clumpy from the scale of galaxies (< 100 kpc) up to superclusters of galaxies ($\gtrsim 100$ Mpc), but the distribution on scales $\gg 100$ Mpc is probably close to uniformity. At the same time, the isotropy of the microwave background and the close agreement of the light element abundance with theoretical standard big bang yields also support the view that the Universe is probably homogeneous and isotropic on large scales from the present back to a redshift $z \sim 10^9$. From the time of decoupling ($z \sim 1400$) up to the present, the evolution of clumps is basically governed only by

Newtonian gravity. So it is in principle unambiguous to extrapolate the present state of clumpiness back to the decoupling era, provided one has sufficiently large computer(s) to take care of the many-body long range interactions. This is in fact being done by many groups [9–10]. So what we really need to explain, from a fundamental point of view, is the origin of density inhomogeneities present *at the end of decoupling*. There are at least three pieces of information that have a bearing on the nature of these inhomogeneities:

(a)   The galaxy covariance function $\xi \equiv \langle (\partial\rho/\rho)(\mathbf{x})(\delta\rho/\rho)(\mathbf{x}+\mathbf{r}) \rangle$ which has a value

$$\xi_0 \cong (r/9.4h^{-1}\,\mathrm{Mpc})^{-1.77} \tag{1}$$

today over the scale 100 kpc $< r <$ 20 Mpc is most consistent with a simple power law continuum density correlation at decoupling of the form

$$\xi_{\mathrm{rec}} \propto r^{-n}, \qquad n \approx 2\text{–}3$$

or equivalently, a root-mean-density contrast spectrum of the form

$$\frac{\delta\rho}{\rho_{\mathrm{rec}}} \propto M^{-\alpha}, \qquad \alpha \approx \tfrac{1}{3} - \tfrac{1}{2}. \tag{2}$$

(Here and in the following $h \equiv H_0/50$ km/sec Mpc). The precise value of α is sensitive to the value of $\Omega_0 (\equiv \rho$ now$/\rho$ crit). It seems likely that α is closer to $\tfrac{1}{3}$ if the Universe is rather open ($\Omega_0 \sim 0.1$) [1]. But if $\Omega_0 \sim 1$ then Davis and Peebles [9] argue that α should be close to $\tfrac{1}{2}$ (white noise).

(b)   The luminosity function of clusters of galaxies is probably consistent with a mass function that evolved from an initial density spectrum of the form $(\delta\rho/\rho_{\mathrm{rec}}) \propto M^{-\alpha}$ with $\alpha \approx \tfrac{1}{3}$ provided certain evolutionary effects are taken into account [1].

(c)   The formation of galaxies at $z \gtrsim 1$–10 means that galactic scale fluctuations reach nonlinearity ($\delta\rho/\rho \sim 1$) at $z \sim 1$–10. Using simple Newtonian perturbation arguments, one can easily show that $\delta\rho/\rho \propto 1/(1+z)$ so long as $\delta\rho/\rho \lesssim 1, (1+z) > \Omega_0^{-1}$. Hence we must have

$$\delta\rho/\rho_{\mathrm{rec}} \sim 1\% - 0.1\% \tag{3}$$

on galactic scale.

The above three results are generally consistent with (but probably do not demand) a postdecoupling density fluctuation spectrum of the form

$$\delta\rho/\rho_{\mathrm{rec}} \sim (M_0/M)^{\alpha}, \qquad M_0 \sim 10^5 - 10^6 M_\odot, \qquad \alpha \approx \tfrac{1}{3} - \tfrac{1}{2}. \tag{4}$$

The value of M_0 is tentalizingly close to the decoupling Jeans mass $M_{\mathrm{Jm}} \sim 10^5 M_\odot \Omega_0^{-1/2}$ to make it suggestive. But I will postpone the discussion of this point until Section V. The correlation function equivalent of Eq. (4) has the form

$$\xi_{rec}(r) \sim (r_0/r)^n, \qquad n \approx 2\text{--}3. \tag{5}$$

The two most significant features of Eqs. (4) or (5) are (i) that fluctuations are strongly correlated $[\xi_{rec} \sim 0(1)]$ on, say, scales of globular clusters or dwarf galaxies, implying that any attempt to use perturbation treatment would fail; (ii) the scale-invariance of the power law, which, superficially at least, resembles the long-range behavior of fluctuation correlation of ordinary matter at critical point, but could also be related to some sort of nonlinear hydrodynamic process (e.g., turbulence spectrum).

III. Linear Perturbations of Uniform Models

Most of the works on galaxy formation in the last two decades were concerned with linear perturbations of uniformly expanding fluid models (Friedmann, Tolman, etc). We shall only give a brief summary here to motivate subsequent discussions. Details can be found in Jones [2], Rees [3], and Silk [5].

The behavior of density perturbations after the decoupling is trivial and actually obtainable using only simple Newtonian arguments in a dust universe. The density contrast has the time dependence:

$$\delta\rho/\rho(t > t_{rec}) \sim At^{-1} + Bt^{2/3}. \tag{6}$$

In galaxy formation and clustering we are only interested in the relatively growing B-mode. One can visualize the $t^{2/3}$ power law growth as due to the lack of an intrinsic time-scale in an expanding universe. If the growth had been exponential: $\delta\rho/\rho \propto e^{at}$, a^{-1} must be some intrinsic time-scale. In a dust universe the only one is $1/\sqrt{G\rho} \sim H^{-1} = \frac{3}{2}t$. Hence $\delta\rho/\rho \propto e^{\int H dt} = t^{2/3}$.

The situation before decoupling is quite complicated because of the effects of radiation pressure and the imperfect coupling between electrons and photons (dissipative effects). Roughly speaking, at least in the linear limit, density perturbations fall into two categories: isothermal and adiabatic modes. Adiabatic modes are just radiation-pressure-driven acoustic waves for wavelengths $\lambda < \lambda_J$, the Jeans length. They damp exponentially whenever λ gets below the damping length $\lambda_D \equiv (\sqrt{\lambda_1}ct)^{1/2}$ ($\lambda_1 \equiv$ scale at which Thomson scattering depth $\tau_{es} \sim 1$) because of dissipative effects. For $\lambda > \lambda_J$ they grow or decay as some power law much like pressureless fluid but the exponent depends on the gauge choice. On the other hand, isothermal modes do not involve the radiation field at all, which remains uniform. The baryons are just "sprinkled" inhomogeneously in the uniformly expanding radiation.

Because radiation drag $\gg$ self-gravity of the baryon clumps, the baryons are forced to coexpand with the radiation and $\delta\rho/\rho_{\text{baryons}} \sim$ constant till decoupling as long as $\lambda \geq \lambda_{\text{Jm}}$, the isothermal Jeans length [11]. For $\lambda < \lambda_{\text{Jm}}$ they are also wiped out by dissipation. The evolution of the critical mass scales M_J, M_D, M_1, M_{Jm} as functions of time in the fireball phase are sketched in Fig. 1. M_J peaks at $\sim 10^{17}\Omega_0^{-2} M_\odot$ in the matter-dominated era before decoupling (provided $\Omega_0 > 0.14$), while M_{Jm} stays constant at $\sim 10^5 \Omega_0^{-1/2} M_\odot$ before decoupling. M_J, of course, drops to M_{Jm} after decoupling. M_D reaches a maximum of $\sim 10^{13}\Omega_0^{-5/4} M_\odot$ just before decoupling [2], while M_1 obviously diverges at decoupling. The principle result is a negative one for adiabatic origins of galaxies: clumps with masses $< 10^{13}\Omega_0^{-5/4} M_\odot$ could not have survived the damping in the fireball phase.

In addition to density perturbations, there could also have been pure vorticity (velocity) or gravity-wave (metric) perturbations without associated density perturbations in the linear limit. These can in turn serve as sources of density modes at higher order. One such scenario is the cosmological analog

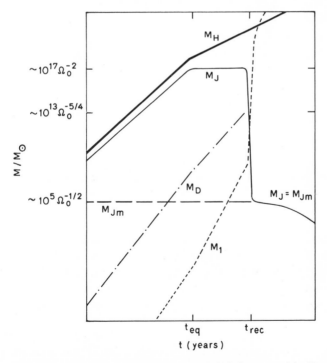

Fig. 1 Evolution of the various critical mass scales in Friedmann models. M_J = radiation (adiabatic) Jeans mass. M_D = damping mass. M_1 = mass corresponding to unit electron scattering depth. M_{Jm} = baryon (isothermal) Jeans mass [3].

of the Lighthill mechanism [12] in which vortical turbulence generates acoustic noise (see, e.g. [13]). The effectiveness of such mechanism is controversial since the turbulence might decay before they become sonic or supersonic. Similar things can be said about gravity-wave perturbations. In any case, fully nonlinear analyses are unavailable and any result based on perturbation arguments is inconclusive (see Section IV).

One thing remains to be discussed: the matching of density perturbation amplitudes across the decoupling era. If the fluid approximation is indeed valid, the matching is trivially done by requiring $\delta\rho/\rho$ and $d/dt(\delta\rho/\rho)$ be continuous across t_{rec} since $\delta\rho/\rho$ satisfies second-order equations before and after decoupling. This procedure gives the postdecoupling B-mode [cf. Eq. (6)] amplitude $(\delta\rho/\rho)_{t_{\text{rec}}^+}$ in terms of the predecoupling acoustic amplitude $(\delta\rho/\rho)_{t_{\text{rec}}^-}$ as

$$\frac{\delta\rho}{\rho_{t_{\text{rec}}^+}} \sim \frac{\delta\rho}{\rho_{t_{\text{rec}}^-}} \left(\frac{M_{J-}}{M}\right)^{1/3}; \qquad M \ll M_{J-}, \qquad (7)$$

where M_{J-} is Jeans mass just before decoupling. However, the big problem (or in fact, the big hope) is that the primeval "fluid" may be very different from a fluid at decoupling and highly nonlinear many-body collective processes may have been triggered by the sharp drop of opacity and effective pressure. Such collective effects should be seriously investigated as possible candidate for the origin of $\delta\rho/\rho$ itself (Section V).

IV. Nonlinear Inhomogeneities in Fireball Phase

In contrast to linear perturbation studies of uniformly expanding models, little work has been done on nonlinear dynamics until recently. Peebles, in a pioneering paper in 1970 [14], first pointed out that just as propagation of ordinary acoustic waves is nonlinearly limited by shock formation, nonlinear adiabatic inhomogeneities in the early Universe would also be limited by shocks. This question has now been investigated in detail by various authors [2,15]. One elegant, analytical way to study this aspect of nonlinear hydrodynamics is via (one-dimensional) simple waves. Special relativistic simple waves solution was first discovered by Taub [16], and was recently generalized by myself to allow for an expanding background [15]. The damping rate of the nonlinear wave, after shock formation, has also been numerically computed [17] (an example is sketched in Fig. 2). When the shock amplitude is weak, it asymptotically decays according to the self-similar law

$$\Delta v \sim \lambda/t, \qquad (8)$$

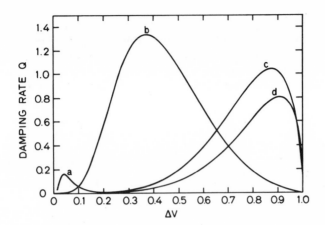

Fig. 2 Dimensionless damping rate $Q \equiv (\lambda/c)|(1/d\Delta v/dt)|$ of relativistic shocked acoustic pulse of initial length λ plotted as functions of initial velocity amplitude Δv for a fluid of γ (adiabatic index) $= \frac{4}{3}$ and sound speed $c_0 =$ (a) $0.01c$, (b) $= 0.1c$, (c) $= 0.5c$ (d) $= 1/\sqrt{3}c$ [17].

where Δv is the velocity amplitude and λ is the wavelength. This self-similar behavior is obtained both for isolated pulses as well as for infinite wave trains. Equation (8) implies

$$\frac{\delta\rho}{\rho} \sim \frac{\Delta v}{c_s} \sim \frac{\lambda}{c_s t} \sim \frac{\lambda}{\lambda_J} \sim \left(\frac{M}{M_J}\right)^{1/3}; \qquad M \ll M_J. \qquad (9)$$

This result agrees with that of Peebles using nonlinear perturbation techniques. It says that large clumps survive nonlinear damping better than small ones. When combined with subsequent photon viscosity and diffusion effects, the final amplitude to emerge at decoupling has the form

$$\frac{\delta\rho^{NL}}{\rho_{rec}} \sim \left(\frac{M}{M_J}\right)^{1/3} e^{-(M/M_D)^{2/3}}, \qquad (10)$$

where M_D is damping mass. This implies that $\delta\rho^{NL}/\rho_{rec} \lesssim 0.01$ for $M \lesssim 10^{12} M_\odot$ and small clumps would not have sufficient amplitude to grow into present-day clumps *even if they had started out nonlinear*. Equations (9) and (10) are, of course, not applicable for $M \gg 10^{12} M_\odot$ since such clumps have sufficiently small amplitude in the fireball phase that shock formation never took place.

The shock formation and dissipation picture is valid so long as the self-gravity of the nonlinear clumps is ignored. But it is obvious that self-gravity must play a role in any nonlinear dynamics. It has nowadays become popular thinking that in a variety of nonlinear systems, nonlinear stability selects out specialized final states possessing a unique spectrum. Solitons [8], convective cells, or even turbulence are probably all examples of such universal selective survival phenomena. Since gravity is obviously dispersive in the Jeans sense, one is tempted to ask: Are there analogs of solitons in a nonlinear self-gravitating fluid? Could the unique mass spectrum of clumps we see be the result of nonlinear selection effects? I have recently taken the first step to address this question and the preliminary answer appears to be "no." The idealized class of model I studied assumes Newtonian gravity, a static Jeans background, and one-dimensional plane waves. In this special case exact nonlinear steady profile waves are found [19], but they are periodic infinite wave trains that admit no solitary wave limit. These nonshocking subsonic waves have no unique spectrum, but their amplitude is limited by

$$\frac{\delta\rho_+}{\rho} \lesssim \left(\frac{M}{M_J}\right)^{2/3}. \tag{11}$$

Above this amplitude no steady profile wave exists. Apparently self-gravity can "hold back" the trend to develop shock only for sufficiently long wavelength, low amplitude perturbations. Figure 3 illustrates a typical wave form. Equation (11) is derived for a special idealized case, it seems intuitive that any nonlinear limit (on the amplitude) obtained with the help of self-gravity must increase with mass scale. Naively, limit (11) can be understood on simple dimensional grounds. The Jeans dispersion relation for linear perturbations is

$$\omega^2 = c_s^2 k^2 - 4\pi G\rho \leftrightarrow U^2 \equiv \frac{\omega^2}{k^2} = c_s^2 - 4\pi G\rho k^{-2}, \qquad c_s = \text{sound speed.}$$

But the effective nonlinear propagation velocity in the absence of gravity is $c_s + v$ instead of c_s where v is velocity amplitude. Using this we have to lowest order in v/c_s:

$$U^2 \cong c_s^2 + 2vc_s - 4\pi G\rho k^{-2} + O(v^2/c_s^2).$$

To make the phase velocity U constant we must therefore demand

$$\frac{\delta\rho}{\rho} \cong \frac{v}{c_s} \sim \frac{2\pi G\rho k^{-2}}{c_s^2} \sim \frac{k_J^2}{k^2}, \qquad k_J \equiv \text{Jeans length}$$

in agreement with Eq. (11).

To summarize, it appears that nonlinear limits obtained both from shock dissipation and gravitational dispersion tend to produce an increasing mass spectrum instead of the decreasing one observed in nature. Whether or not other nonlinear effects could be responsible for a decreasing mass spectrum

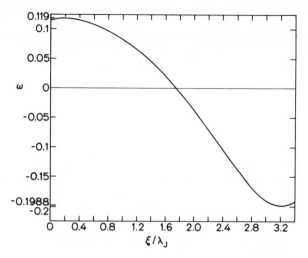

Fig. 3 Waveform of a nonlinear periodic wave in a $\gamma = \frac{4}{3}$ static Jeans universe. $\mathcal{M} \equiv U/c_s = 1/\sqrt{2}$. A is some constant related to the amplitude ($A = 0.005$). $\lambda = 6\lambda_J$. Note that the crest is flatter and broader than the trough. This asymmetry due to nonlinearity becomes more pronounced as $\mathcal{M}$ is decreased or A increased [19].

remains to be investigated. For example, nonlinear coupling between adiabatic and isothermal modes in a two-fluid approximation might play an important role and is definitely worth investigating. However, it does seem unlikely that a decreasing final mass spectrum can be consistent with completely "chaotic" initial conditions with nonlinear fluctuations occurring on all scales. This is because the largest clumps enter the horizon the latest and have the longest time available to freely condense. We must therefore somehow start with sufficiently "mild" conditions on large scales. This makes the "chaotic" picture unattractive on fundamental grounds.

V.   Hints of a Gravithermal Origin

Intuitively, a smooth monotonically decreasing mass spectrum strongly suggests that the clumping is the result of some causal physical process that increases its domain of influence with time, rather than some ad hoc initial conditions, because the domains of communication grow with time. Empirically, the amplitude of spectrum (4) makes it tempting to associate the amplitude-fixing mass M_0 with the decoupling Jeans mass M_{Jm}

$$M_{\text{Jm}} \equiv \frac{4\pi}{3} mn \left(\frac{\pi kT}{Gm^2n}\right)^{3/2} \sim 10^5 \Omega_0^{1/2} M_\odot , \qquad (12)$$

where m is the baryon mass, T the temperature, and n the baryon density. In that case the density correlation function has the amplitude

$$\xi_{rec}(r) \sim (r_{Jm}/r)^n, \quad n \approx 2; \quad r \gg r_{Jm} \equiv \text{Jeans length}. \tag{13}$$

If this conjecture is correct, it may be the most important clue to the nature and origin of the primeval inhomogeneities. It would suggest that the finite density fluctuations have a gravithermal origin.

The most attractive candidate for the spontaneous generation of such density correlations would appear to be the decoupling process itself. Note that $r_{Jm} \sim c_s t_{rec}$ is approximately the maximum distance a thermal baryon could travel in one expansion time at decoupling, i.e., the maximum correlation length peculiar velocity can build up. A simple *hypothetical* mechanism can in fact be exhibited to generate the density fluctuations spectrum of the form (4) or (13). If the baryons inside a comoving, expanding sphere of radius r somehow manage to have all their thermal velocities of magnitude c_s aligned *against* their radial expansion velocity, then the comoving sphere will expand more slowly than its surroundings and, in approximately one expansion time, will develop a density contrast of

$$\frac{\delta\rho}{\rho} = 3\frac{\delta r}{r} \sim \frac{3c_s t_{rec}}{r} \sim \frac{r_{Jm}}{r} \sim \left(\frac{M_{Jm}}{M}\right)^{1/3}. \tag{14}$$

Statistically speaking, what one needs to generate density correlation (13) is to have all the baryon random velocity directions correlated macroscopically *on all relevant scales*. But what could have caused such velocity correlation to develop? At present we have little idea, but suspect that it might be related to some sort of first-order transition triggered by the decoupling process. After all, the transition from opaqueness to transparency at decoupling is very sharp and the effective pressure drop almost discontinuously by nine orders of magnitude in a few expansion times. One analogy that readily comes to mind is the spontaneous development of domains in ferromagnetic transitions below critical temperature, which occurs on a diffusion time scale. Here something might be responsible for the development of "velocity domains." Since photon is the only agent that can traverse all the relevant scales in approximately one expansion time at decoupling, it seems plausible that the baryon velocity correlations might be triggered by the last scattering with photons. Much more work is needed before we can have any definite opinion about fully nonlinear collective processes that might occur at decoupling. But one thing is certain: The conventional hydrodynamic picture of a smooth transition from collision-dominated fluid to collisionless dust is inadequate. Our hope of getting something from nothing precisely hinges on this breakdown of the simple fluid (or even kinetic) approximation.

The scale-invariance of correlation (13) may just be a reflection of the long-range nature of gravity. But it could also be another clue to the nature of the fluctuations, since long-range correlation of fluctuations is obtained for ordinary matter only at critical point (see, e.g. [20]). Recently, Cardy (personal communication) has sucessfully shown by renormalization group arguments that an equilibrium gravitational pseudo-plasma (with equal amounts of positive and negative charges) possesses critical point(s), at least in the lattice approximation. The primeval density inhomogeneities, with the uniformly expanding background subtracted off, may be visualized as a pseudo-plasma. There is of course no reason to believe equilibrium arguments have any meaning for a dynamical, expanding universe. But such results may be useful as a basis for formulating a dynamical theory of large density correlations in a self-gravitating gas.

VI. Summary

In this paper I have discussed two alternative possibilities for the origin of the unique spectrum of cosmic density inhomogeneities before or at decoupling: nonlinear selective survival versus gravithermal transition at decoupling. Our current knowledge about nonlinear hydrodynamic limits does not favor the selective survival conjecture; but the study of first-order transition-type collective processes at decoupling is still at its infant stage. It is too early to tell if it is a viable mechanism, but at least there might be a hope.

References

[1] Gott, J. R., III, *Annu. Rev. Astron. Astrophys.* **15**, 235 (1977), and references therein.
[2] Jones, B. J. T., *Rev. Mod. Phys.* **48**, 107 (1976).
[3] Rees, M. J., *Proc. Int. Sch. Phys.* "Enrico Fermi" (R. Sachs, ed.), Vol. 47. Academic Press, New York, 1971.
[4] Field, G. B., in "Stars and Stellar Systems" (A. Sandage, M. Sandage, and J. Kristian, eds.), Vol. 9, p. 359. Univ. of Chicago Press, Chicago, Illinois, 1975.
[5] Silk, J., in "Confrontation of Cosmological Theories with Observational Data" (M. S. Longair, ed.), p. 175. Reidel Publ., Dordrecht, Netherlands, 1974.
[6] Peebles, P. J. E., "Physical Cosmology." Princeton Univ. Press, Princeton, New Jersey, 1971.
[7] Weinberg, S., "Gravitation and Cosmology." Wiley (Interscience), New York, 1972.
[8] Liang, E. P. T., and Sachs, R. K., in "Einstein Centennial Volume" (A. Held and P. Bergmann, eds.), Vol. II, p. 329. Plenum Press, New York, 1979.
[9] Davis, M., and Peebles, P. J. E., *Astrophys. J., Suppl. Ser.* **34**, 425 (1977).
[10] Aarseth, S. J., Gott, J. R., and Turner, E. L., in *Int. Ast. Union Symp. 79*, (M. Longair and J. Einsats, eds.), Reidel Publ., Dordrecht, Netherlands, 1978; see also Fall, S. M., *Rev. Mod. Phys.* **51**, 21 (1979); *Mon. Not. R. Astron. Soc.* **185**, 165 (1978).

[11]   Peebles, P. J. E., *Astrophys. J.* **142**, 1317 (1965).

[12]   Lighthill, M. J., *Proc. R. Soc. London, Ser. A* **267**, 147 (1962).

[13]   Silk, J., and Ames, S., *Astrophys. J.* **178**, 77 (1972); Ozernoi, L. M., and Chebyshev, G. V.,
        Sov. Astron. A. J. **14**, 615 (1971), and references therein.

[14]   Peebles, P. J. E., *Phys. Rev. D* **1**, 397 (1970).

[15]   Liang, E. P. T., *Astrophys. J.* **204**, 235 (1976); **211**, 361 (1977); **216**, 206 (1977).

[16]   Taub, A. H., *Phys. Rev.* **14**, 328 (1948).

[17]   Liang, E. P. T., *Phys. Rev. Lett.* **39**, 191 (1977).

[18]   Scott, A. G., Chu, F. Y. F., and McLaughlin, D. W., *Proc. IEEE* **61**, 1443 (1973).

[19]   Liang, E. P. T., *Astrophys. J.* **230**, 325 (1979).

[20]   Ma, S. K., "Modern Theory of Critical Phenomena." Benjamin, New York, 1976.

14

Automorphisms of Formal Algebras Associated by Deformation with a Symplectic Manifold

André Lichnerowicz

Collège de France
Paris, France

Abstract

It is possible to give a complete description of Classical Mechanics in terms of sympletic geometry and Poisson brackets. It is the essence of the Hamiltonian formalism. *Deformations* of the Poisson Lie algebra and deformations of the associative algebra defined by the usual product of scalar functions determine interesting formal algebras and give a new approach for quantum mechanics [2].

The general study of the automorphisms of such algebras gives precious information on the *invariance problems concerning quantum mechanics*. In this paper, my purpose is to show the main points of such a study. I consider here only dynamical systems with a finite number of degrees of freedom, but the approach and a significant part of the results can be extended to physical fields.

I.   Classical Dynamics and Symplectic Manifolds

(a)   Consider a dynamical system with time-independent constraints and n degrees of freedom. The corresponding configuration space is an arbitrary differentiable manifold M of dimension n. It is well known that the cotangent bundle T^*M admits a natural symplectic structure defined by the exact Liouville two-form. For the Hamiltonian formalism, a dynamical state of

203

the system is nothing other than a point of $W = T^*M$, which corresponds to the usual *phase space*. The analysis of the equations of mechanics has shown that it is essential to introduce changes of the classical variables (q^α, p_α) $(\alpha = 1, \ldots, n)$, which do not respect the cotangent structure of W. We are thus led to introduce as *phase space* a symplectic manifold (W, F) of dimension $2n$ and fundamental two-form F. All the elements are supposed C^∞-smooth. We denote by $\mu\colon TW \to T^*W$ the isomorphism of vector bundles defined by $\mu(X) = -i(X)F$, where $i(\ )$ is the interior product; this isomorphism can be extended to tensors in a natural way. We denote by Λ the skewsymmetrical contravariant two-tensor $\mu^{-1}(F)$. We put $N = C^\infty$ $(W; \mathbb{R})$.

A *symplectic vector field* is a vector field X such that $\mathscr{L}(X)F = 0$ (where $\mathscr{L}$ is the Lie derivative); it defines an infinitesimal automorphism of the structure. We denote by L the (infinite-dimensional) Lie algebra of the symplectic vector fields; $X \in L$ iff the one-form $\mu(X)$ is closed. If $X, Y \in L$

$$\mu([X, Y]) = di(\Lambda)(\mu(X) \wedge \mu(Y)). \tag{I.1}$$

Let L^* be the subspace of L defined by the Hamiltonian vector fields, inverse images of the exact one-forms $(X_u = \mu^{-1}(du); u \in N)$; L^* is exactly the commutator ideal $[L, L]$ of L (each element of $[L, L]$ being a finite sum of brackets of elements of L). If $\bar{N}$ is the space of the classes of elements of N modulo the additive constants, the natural isomorphism between the spaces L^* and $\bar{N}$ induces on $\bar{N}$ a Lie algebra structure defined in the following way: If $\bar{u}, \bar{v} \in \bar{N}$, it follows from (I.1) that the function $w = i(\Lambda)(d\bar{u} \wedge d\bar{v})$ defines a class that is the bracket of $\bar{u}, \bar{v}$. The function w is the Poisson bracket of $\bar{u}, \bar{v}$, or of two representatives u, v in N. We put

$$\{u, v\} = P(u, v) = i(\Lambda)(du \wedge dv) = \mathscr{L}(X_u)v. \tag{I.2}$$

The Poisson bracket P defines a Lie algebra structure and we have a homomorphism of the Poisson Lie algebra (N, P) on the Lie algebra L^* of the Hamiltonian vector fields.

(b)   The geometry of the dynamical system being given by the symplectic manifold (W, F), the dynamics are determined by a function $H \in N$, the Hamiltonian of the system, which defines a Hamiltonian vector field X_H. A *motion of the dynamical system is given, by definition, by an integral curve $c(t)$ of the Hamiltonian vector field X_H, the parameter t being the time.* Such is the geometrical meaning of the classical equations of Hamilton. We can adopt another viewpoint. The space N admits the following two algebraic structures:

(1)   an *associative algebra* structure defined by the usual product of functions (the algebra is here commutative); and

(2)   a *Lie algebra* structure defined by the Poisson bracket.

The derivations of the product being given by the vector fields, we have

$$\{w, uv\} = \{w, u\} \cdot v + u \cdot \{w, v\}, \qquad (u, v, w \in N). \tag{I.3}$$

Consider a family u_t of elements of N satisfying the differential equation

$$du_t dt = \{H, u_t\} \tag{I.4}$$

and taking the initial value u_0 at $t = 0$. It follows from (I.3) that the evolution in time of u_t follows the trajectories that appear in the first viewpoint; (I.4) can be considered as the intrinsic equation of classical dynamics.

(c)   We can completely describe classical mechanics in terms of the two laws of composition of N, connected by (I.3). It is natural to study whether it is possible to deform in a suitable way these two algebraic laws, so that we obtain a model isomorphic to the usual quantum mechanics. The first results obtained by Bayen *et al.* [2] and Vey [8] are positive.

II.   The Formal Algebras

(a)   Let $E(N; v)$ be the space of the formal functions of $v \in \mathbb{C}$ with coefficients in N. A formal differential deformation of the associative algebra $(N, \cdot)$ is given by a bilinear map $N \times N \to E(N; v)$, where v is the deformation parameter. Let

$$u*_v v = \sum_{r=0}^{\infty} v^r C^r(u, v) = (uv) + \sum_{r=0}^{\infty} v^r C^r(u, v), \qquad (u, v \in N) \tag{II.1}$$

which satisfies formally the associativity identity

$$(u*_v v)*_v w = u*_v(v*_v w), \qquad (u, v, w \in N). \tag{II.2}$$

The $C^r: N \times N \to N$ are (for $r \geq 1$) differential bilinear maps; $*_v$ defines on $E(N; v)$ an associative algebra structure called a *formal associative algebra* obtained by deformation. We consider only the formal associative algebras satisfying the following two assumptions:

Assumptions (H)   (1) $C^r(u, v)$ is symmetric in u, v if r is even, skewsymmetric if r is odd. (2) $C^1 = P$ and C^r is null on the constants.

(b)   In a similar way, a formal differential deformation of the Poisson Lie algebra (N, P) is given by an alternating bilinear map $N \times N \to E(N; \lambda)$, where $\lambda \in \mathbb{C}$ is the deformation parameter; let

$$[u, v]_\lambda = \sum_{r=0}^{\infty} \lambda^r C^{2r+1}(u, v) = P(u, v) + \sum_{r=1}^{\infty} \lambda^r C^{2r+1}(u, v), \tag{II.3}$$

which satisfies formally Jacobi's identity. The C^{2r+1}: $N \times N \to N$ are (for $r \geq 1$) alternating differential bilinear maps; (II.3) defines on $E(N; \lambda)$ a Lie algebra structure called a *formal Lie algebra* obtained by deformation of the Poisson Lie algebra. We consider here only Lie algebras such that *the C^{2r+1} are null on the constants.*

A formal associative algebra satisfying the assumptions (H) generates by skewsymmetrization a formal Lie algebra satisfying the assumption

$$[u, v]_\lambda = (2v)^{-1}(u *_v v - v *_v u) \qquad (\lambda = v^2).\qquad\qquad \text{(II.4)}$$

III. Deformations and Cohomology

I will first recall briefly and extend the main elements of the theory of Gerstenhaber [5] concerning the deformations of the algebraic structures.

(a)   Let W be a differentiable manifold. If $N = C^\infty(W; \mathbb{R})$, $(N, \ )$ is the associative algebra defined by the usual product of functions. We consider the associative deformations of this algebra. Derivations and deformations arise from the same cohomology, the so-called *Hochschild cohomology* of $(N, .)$. This cohomology is defined in the following way: a p-cochain of $(N, .)$ is a p-linear map of N^p into N, the 0-cochains being identified with the elements of N. The coboundary of the p-cochain C is the $(p + 1)$-cochain $\tilde{\partial}C$ defined by

$$\begin{aligned}
\tilde{\partial}C(u_0, \ldots, u_p) = {} & u_0 C(u_1, \ldots, u_p) - C(u_0 u_1, u_2, \ldots, u_p) \\
& + C(u_0, u_1 u_2, \ldots, u_p) + \cdots + (-1)^p C(u_0, \ldots, u_{p-1} u_p) \\
& + (-1)^{p+1} C(u_0, \ldots, u_{p-1}) u_p,
\end{aligned}\qquad \text{(III.1)}$$

where $u_\lambda \in N$. A one-cocycle of $(N, .)$ is a derivation of this algebra. A p-cochain C is called *local* if, for each $u_\lambda \in N$ such that $u_{\lambda|U} = 0$ on a domain U, we have $C(u_1, \ldots, u_\lambda, \ldots, u_p)_{|U} = 0$. If C is local, $\tilde{\partial}C$ is local. A p-cochain C is called *d-differential* $(d \geq 0)$ if it is defined by a multi-differential operator of maximum order d in each argument. If C is a $(d + 1)$ differential one-cochain, $\tilde{\partial}C$ is d-differential. It is necessary to study the endomorphisms T of N such that $\tilde{\partial}T$ is d-differential. We have

$$\tilde{\partial}T(u, v) = u \cdot Tv + vTu - T(uv).$$

It follows that if T is an endomorphism such that $\tilde{\partial}T$ is *local*, T is local itself. From this locality, it follows by a nontrivial argument [6]:

Proposition   If T is an endomorphism of N such that $\tilde{\partial}T$ is a d-differential $(d \geq 0)$ two-cochain, T is $(d + 1)$-differential itself. If $\tilde{\partial}T$ is null on the constants, it is the same for T.

(b)   Consider a bilinear map $N \times N \to E(N; v)$, which gives rise to a formal series in v

$$u*_v v = \sum_{r=0}^{\infty} v^r C^r(u, v) = uv + \sum_{r=1}^{\infty} v^r C^r(u, v), \qquad (\text{III.2})$$

where the C^rs $(r \geq 1)$ are differential two-cochains. Introduce the three-cochains

$$E^t(u, v, w) = \sum_{r+s=t} C^r(C^s(u, v), w) - \sum_{r+s=t} C^r(u, C^s(v, w)), \quad (r, s \geq 1). \quad (\text{III.3})$$

The associativity identity (II.2) is satisfied iff $\tilde{\partial} C^1 = 0$ and, for each $t \geq 2$, $\tilde{\partial} C^t = E^t$. We note that $\tilde{\partial} P = 0$ and that our choice $C^1 = P$ is consistent. If (III.2) is limited to the order q, we have *a deformation of order q* if the associativity identity is satisfied up to order $(q + 1)$. If such is the case, E^{q+1} is automatically a three-cocycle of $(N, .)$. We can find a two-cochain C^{q+1} such that $\tilde{\partial} C^{q+1} = E^{q+1}$ iff E^{q+1} is exact; E^{q+1} defines a cohomology class that is *the obstruction at order $(q + 1)$* to the construction of a deformation.

Consider a formal series in v:

$$T_v = \sum_{s=0}^{\infty} v^s T_s = Id + \sum_{s=1}^{\infty} v^s T_s, \qquad (\text{III.4})$$

where the T_ss $(s \geq 1)$ are endomorphisms of N; T_v acts naturally on $E(N; v)$. Consider also another bilinear map $N \times N \to E(N; v)$ corresponding to the formal series

$$u*'_v v = uv + \sum_{r=1}^{\infty} v^r C'^r(u, v), \qquad (\text{III.5})$$

where the C'^rs are differential two-cochains again. Suppose (III.4) and (III.5) are such that we have formally the identity:

$$T_v(u*'_v v) = T_v u *_v T_v v. \qquad (\text{III.6})$$

This identity can be translated to

$$C'^t - C^t + G^t = \tilde{\partial} T_t, \qquad (t = 1, 2, \ldots), \qquad (\text{III.7})$$

where $G^1 = 0$ and

$$G^t(u, v) = \sum_{r+s=t} T_s C'^r(u, v) - \sum_{s+s'=t} T_s u, T_{s'} v - \sum_{r+s=t} (C^r(T_s u, v)$$
$$+ C^r(u, T_s v)) - \sum_{r+s+s'=t} C^r(T_s u, T_{s'} v), \qquad (r, s, s' \geq 1). \quad (\text{III.8})$$

It follows from the proposition of (a) that *the $T_s s$ ($s \geq 1$) are necessarily differential operators*. Using universal formulas, we prove by recursion the following:

Proposition   The associative deformation (III.2) *of* $(N, .)$ *being given, each formal series* (III.4), *where the T_s ($s \geq 1$) are differential operators, generates a unique bilinear map* (III.5) *satisfying* (III.6). *This map is a new associative deformation which is called equivalent to* (III.2). *In particular a deformation is called trivial if it is equivalent to the identity deformation* ($C^r = 0$ *for every* $r \geq 1$).

If two deformations are equivalent at order q, there appears a two-cocycle $(C'^{q+1} - C^{q+1} + G^{q+1})$, which is *the obstruction to equivalence* for order $(q + 1)$. In particular, an infinitesimal deformation (deformation of order 1) is trivial iff the two-cocycle C^1 defining it is exact. For a symplectic manifold, the two-cocycle P is certainly *nonexact*. It follows that the associative deformations satisfying the assumptions (H) are nontrivial even for the order 1.

(c)   We have similar definitions and results for the deformations of the Poisson Lie algebra (N, P) of a symplectic manifold. There is a cohomology, the so-called *Chevalley cohomology* [3] of a Lie algebra, which plays the same role for the Lie algebra deformations as the Hochschild cohomology for the associative deformations. A p-cochain of (N, P) is an alternating p-linear map of N^p into N; the coboundary of the p-cochain C is the $(p + 1)$-cochain ∂C defined by

$$\partial C(u_0, \ldots, u_p) = \varepsilon_0^{\lambda_0 \cdots \lambda_p} \left(\frac{1}{p!} \{ u_{\lambda_0}, C(u_{\lambda_1}, \ldots, u_{\lambda_p}) \} \right.$$

$$\left. - \frac{1}{2(p-1)!} C(\{u_{\lambda_0}, u_{\lambda_1}\}, u_{\lambda_2}, \ldots, u_{\lambda_p}) \right), \quad \text{(III.9)}$$

where $u_\lambda \in N$ and where ε is the skewsymmetric Kronecker indicator. The one-cocycles of (N, P) are the *derivations* of this Lie algebra, the exact one-cocycles are the *inner derivations*. If C is a d-differential ($d \geq 1$) p-cochain ∂C is also d-differential. We have proved [4]:

Lemma 1   Let (W, F) be a noncompact symplectic manifold. If T is an endomorphism of N such that $C = \partial T$ is d-differential ($d > 1$), T is d-differential itself.

The study of the compact case gives [1, 6].

Lemma 2   Let (W, F) be a compact symplectic manifold. If T is an endomorphism of N such that $C = \partial T$ is d-differential $(d \geq 1)$, T has the form:

$$Tu = Qu + C \int_W uF^n,$$

where $C \in \mathbb{R}$ and where Q is a differential operator of order d.

We deduce from these two lemmas the general proposition.

Proposition   If C is an exact d-differential two-cocycle $(d \geq 1)$ of the Poisson Lie algebra (N, P), there is, on the symplectic manifold (W, F), a differential operator T of order d such that $C = \partial T$.

In particular it follows from this proposition and from the lemmas (see [1]):

Corollary   For an arbitrary symplectic manifold, the derivations D of the Poisson Lie algebra (N, P) which are null on the constants are given by $D = \mathscr{L}(X)$, where $X \in L$.

IV.   Vey Algebras

(a)   Let (W, F) be a symplectic manifold. Such a manifold admits atlases of charts for which F (or Λ) have constant components [*natural* charts $\{x^i\}$ $(i, j = 1, \ldots, 2n)$].

A *symplectic connection* Γ is a linear connection without torsion such that $\nabla F = 0$, where ∇ is the operator of covariant differentiation defined by Γ. If $\{\Gamma^i_{jk}\}$ are the usual coefficients of a connection Γ in a natural chart $\{x^i\}$, we introduce the coefficients $\Gamma_{ijk} = F_{i1}\Gamma^1_{jk}$. Such coefficients $\{\Gamma_{ijk}\}$ define a symplectic connection iff they are completely symmetric for every natural chart. A symplectic manifold admits infinitely many symplectic connections; the difference between two symplectic connection is given by a symmetric covariant three tensor.

Introduce the bidifferential operators P^r of maximum order r on each argument defined by the following expression on each domain U of an arbitrary chart $\{x^i\}$:

$$P^r(u, v)_{|U} = \Lambda^{i_1 j_1} \cdots \Lambda^{i_r \, j_r} \nabla_{i_1, \ldots, i_r} u \nabla_{j_1, \ldots, j_r} v. \qquad \text{(IV.1)}$$

We put $P^0(u, v) = u \cdot v$ and we have $P^1 = P$. If the connection Γ is *flat*, it is well known that

$$u *_v v = \sum_{r=0}^{\infty} (v^r/r!)P^r(u, v) = \exp(vP)(u, v) \qquad \text{(IV.2)}$$

defines a deformation of $(N, .)$; (IV.2) is called the Moyal product (see [7,9]).

In the general case, denote by $\mathscr{L}(X_u)\Gamma$ the symmetric covariant three-tensor defined by means of the Lie dimension of the symplectic connection Γ by the Hamiltonian vector field X_u. The two-cochain S_Γ^3 given by

$$S_\Gamma^3(u, v)_{|U} = \Lambda^{i_1 j_1}\Lambda^{i_2 j_2}\Lambda^{i_3 j_3}(\mathscr{L}(X_u)\Gamma)_{i_1 i_2 i_3}(\mathscr{L}(X_u)\Gamma)_{j_1 j_2 j_3} \qquad \text{(IV.3)}$$

admits the same principal symbol as P^3 and is a *nonexact* Chevalley two-cocycle. Its cohomology two-class β depends only upon the symplectic structure of the manifold.

(b)   Introduce now the following notations: We denote by Q^r a bi-differential operator of maximum order r on each argument, null on the constants and such that its principal symbol coincides with the principal symbol of P^r; Q^r is symmetric in u, v if r is even, skewsymmetric if r is odd. We take in particular $Q^0(u, v) = uv$, $Q^1(u, v) = P(u, v)$, $Q^3 \in \beta$. Vey [8] has recently proved the following:

Theorem (Vey)   *Let* (W, F) *be a symplectic manifold such that the third Betti number* $b_3(W)$ *is null. There exist deformations of the Poisson Lie algebra of the manifold such that*

$$[u, v]_\lambda = \sum_{r=0}^{\infty} \left(\frac{\lambda^r}{(2r + 1)!}\right)Q^{2r+1}(u, v). \qquad \text{(IV.4)}$$

(IV.4) defines on $E(N; \lambda)$ a *Vey Lie algebra structure*. General explicit forms for Q^{2r+1} are not known. For the two-cocycle Q^3, there is a unique symplectic connection Γ such that $Q^3 = S_\Gamma^3 + \partial K$, where K is a differential operator of order ≤ 2, such that $K(1) = $ constant.

A *Vey associative algebra* is given on $E(N; v)$ by a deformation of $(N, \ )$ having the form

$$u *_v v = \sum_{r=0}^{\infty} \left(\frac{v^r}{r!}\right)Q^r(u, v). \qquad \text{(IV.5)}$$

The general problem of the existence of such a $*_v$-product is more difficult than the problem solved by Vey and the answer is unknown. Many examples are known in particular for large classes of cotangent bundles of homogeneous spaces.

V.   Derivations and Class of Automorphisms for a Formal Lie Algebra

Let (W, F) be a symplectic manifold *admitting a formal Lie algebra* (II.3) null on the constants.

(a)   Consider a symplectic vector field $Z \in L$; if U is a contractible domain of W, there is $w_u \in C^\infty(U; \mathbb{R})$, defined up to an additive constant, such that

$Z_{|U} = \mu^{-1}(dw_U)$; C^{2r+1} being null on the constants, we define on N a differential operator C_Z^{2r+1} by

$$C_Z^{2r+1}(u)_{|U} = C^{2r+1}(w_U, u_{|U}); \qquad (u \in N). \tag{V.1}$$

Let $E(L; \lambda)$ (respectively, $E(L^*; \lambda)$) be the space of formal functions in λ with coefficients in L (respectively, L^*). These spaces admit natural Lie algebra structures. We can prove by recursion:

Proposition   The Lie algebra of the derivations of a formal Lie algebra (II.3) which are null on the constants is isomorphic to the Lie algebra $E(L; \lambda)$ by the isomorphism $\rho: Z_\lambda = \sum \lambda^s Z_s \in E(L; \lambda) \to D_\lambda$, where

$$D_\lambda = \sum_{r,s=0}^{\infty} \lambda^{r+s} C_Z^{2r+1}. \tag{V.2}$$

The Lie algebra of the inner derivations is isomorphic to $E(L^*; \lambda)$.

(b)   Consider an automorphism of the space $E(N; \lambda)$:

$$A_\lambda = A_0 + \sum_{s=1}^{\infty} \lambda^s A_s, \tag{V.3}$$

where A_0 is an automorphism of the space N and the $A_s(s \geq 1)$ are endomorphisms of N; A_λ is an automorphism of the Lie algebra (II.3) iff:

$$A_\lambda[u, v]_\lambda - [A_\lambda u, A_\lambda v]_\lambda = 0. \tag{V.4}$$

If $A_0 = \text{Id}$, we say that A_λ has *a trivial main part*. Let symp (W, F) be the group of all the symplectomorphisms of (W, F) and $\text{symp}_c(W, F)$ the component connected to the identity by differentiable paths. Consider a symplectomorphism σ isotopic to the identity of W and let $\sigma(t)$ $(0 \leq t \leq 1)$ be a symplectic isotopy such that $\sigma(0) = \text{Id}$, $\sigma(1) = \sigma$. For each t:

$$\dot{\sigma}(t) = \frac{d\sigma(t)}{dt} \sigma(t)^{-1}$$

defines a symplectic vector field; according to (V.5), $\dot{\sigma}(t)$ determines for each t a derivation:

$$\rho(\dot{\sigma}(t)) = D_\lambda(t) = \sum_{r=0}^{\infty} \lambda^r C_{\dot{\sigma}(t)}^{2r+1}.$$

Consider the differential equation

$$dA_\lambda(t)/dt = D_\lambda(t)A_\lambda(t),$$

where $A_\lambda(t)$ is an automorphism of the space $E(N;\lambda)$, and suppose $A_\lambda(0) = \mathrm{Id}$. Studying the solution of this equation, we can construct a one-parameter differentiable family of automorphisms of (II.3) of the form

$$A_\lambda(t) = \sigma^*(t)\left(\mathrm{Id} + \sum_{s=1}^{\infty} \lambda^s V_s(t)\right),$$

where the $V_s(t)$ are differential operators on N, depending differentiably upon t, null on the constants. It follows that

Proposition   For each element σ of $symp_c(W, F)$, there exists an automorphism of the formal Lie algebra (II.3) of the form

$$A_\lambda = \sigma^*\left(\mathrm{Id} + \sum_{s=1}^{\infty} \lambda^s V_s\right) = \left(\mathrm{Id} + \sum_{s=1}^{\infty} \lambda^s B_s\right)\sigma^*, \qquad \text{(V.5)}$$

where the V_s (respectively, B_s) are differential operators on N, null on the constants.

VI.   Derivations of a Formal Associative Algebra

In the following part, we suppose that, for the formal Lie algebra of (W, F), *there is a formal associative algebra* (II.1) *satisfying the assumptions* (H) *which generate* (II.3) *by skewsymmetrization.*

(a)   We study the uniqueness of such a $*_\nu$ product. Suppose that there is a $*'_\nu$ product satisfying the same assumptions and generating the same Lie algebra. We have $C'^{2r+1} = C^{2r+1}$ and we are led to study by recursion $M^{2r} = C'^{2r} - C^{2r}$. If $M^{2r} = 0$ for $0 \le r \le (t-1)$, the identity $\tilde{\partial}C'^{2t+1} = \tilde{\partial}C^{2t+1}$ can be translated by

$$P(M^{2t}(u, v), w) + M^{2t}(P(u, v), w) - P(u, M^{2t}(v, w)) - M^{2t}(u, P(v, w)) = 0.$$

It is possible to show the following [6]:

Lemma   Let M be a bidifferential operator, symmetric in both arguments, null on the constants. The relation

$$P(M(u, v), w) + M(P(u, v), w) - P(u, M(v, w)) - M(u, P(v, w)) = 0$$

implies $M = 0$.

It follows that

Uniqueness theorem   Suppose that there is on (W, F) a formal associative algebra (II.1) *satisfying the assumptions* (H) *and generating the formal Lie algebra* (II.3). *Every associative algebra satisfying the assumptions* (H) *and generating* (II.3) *coincides with* (II.1).

(b)   The Lie algebra structure (II.1) on $E(N; \lambda)$ can be transported to $E(N; v)$, with $\lambda = v^2$:

$$[u, v]_{v^2} = P(u, v) + \sum_{r=1}^{\infty} v^{2r} C^{2r+1}(u, v). \tag{VI.1}$$

It follows from the proposition of Section V,a that the Lie algebra of the derivations $D_v = \sum v^s D_s$ of the Lie algebra (VI.1), which are null on the constants, is isomorphic to the Lie algebra $E(L; v)$ by the isomorphism ρ: $Z_v = \sum v^s Z_s \in E(L; v) \to D_v$, with

$$D_v = \sum_{r, s \geq 0} v^{2r+s} C_{Z_s}^{2r+1}. \tag{VI.2}$$

Each derivation D_v of the associative algebra (II.1) is a derivation of the Lie algebra (VI.1), which is null on the constants, according to the proposition of Section III(a); D_v necessarily has the form (VI.2). Conversely, it is clear that each derivation (VI.2) of the Lie algebra is a derivation of the associative algebra. It follows that

Theorem   The derivations of a formal associative algebra satisfying the assumptions (H) and the derivations of the corresponding Lie algebra that are null on the constants coincide and are given by (VI.2). If $b_1(W) = 0$, the derivations of the associative algebra are all inner derivations.

VII.   Automorphisms of the Associative Algebra and of the Corresponding Lie Algebra

(a)   We study the automorphisms $A_v = A_0 + \sum_{s \geq 1} v^s A_s$ of the associative algebra (II.1). For order 0, we have: $A_0(uv) = A_0 u \cdot A_0 v$. It follows classically that there is a diffeomorphism σ of W such that $A_0 = \sigma^*$. We obtain for order 1:

$$\sigma^* P(u, v) = P(\sigma^* u, \sigma^* v)$$

and σ is necessarily a *symplectomorphism*. We deduce from Section III:

Proposition   Each automorphism A_v of an associative algebra satisfying the assumptions (H) has the form

$$A_v = \left(\mathrm{Id} + \sum_{s=1}^{\infty} v^s B_s \right) \sigma^*, \tag{VII.1}$$

where σ is a symplectomorphism and where the $B_s s$ are differential operators that are null on the constants.

(b)   Consider for a moment the automorphisms A_v of the *Lie algebra* (VI.1), which are *even with respect to v*. We have

$$A_v = \left(\text{Id} + \sum_{s=1}^{\infty} v^{2s}B_{2s}\right)\sigma^*. \qquad (\text{VII.2})$$

The associative product $\hat{*}_v$ defined by

$$A_v(u*_v v) = A_v u \,\hat{*}_v\, A_v v$$

gives by skewsymmetrization the Lie algebra (VI.1). It is easy to see that $\hat{C}^{2r}$ is symmetric in u, v and we can prove by recursion that $\hat{C}^{2r+1}$ is skew-symmetric in u, v and thus equal to C^{2r+1}. It follows from the uniqueness theorem that

Proposition   Consider an associative algebra (II.1) *satisfying the assumptions* (H). *The group of the automorphisms of this algebra that are even in v coincides with the group of the automorphisms of the corresponding Lie algebra that have the form* (VII.2), *where σ is a symplectomorphism and where the B_ss are differential operators that are null on the constants.*

VIII.   Automorphisms of an Associative Algebra which Are Trivial Main Parts

(a)   We study the automorphisms of the associative algebra (II.1) given by

$$A_v = \text{Id} + \sum_{s=1}^{\infty} v^s A_s, \qquad (\text{VIII.1})$$

where the A_s $(s \geq 1)$ are necessarily differential operators that are null on the constants. The group $\text{aut}_t\,(*_v)$ of these automorphisms that have *a trivial main part* is an invariant subgroup of the group $\text{aut}(*_v)$ of all the automorphisms of (II.1).

Let $\hat{E}(N; v)$ be the subset of $E(N; v)$ defined by the elements $a_v = \sum v^s a_s$ such that $a_0 > 0$ on W. Each element a_v of $\hat{E}(N; v)$ admits an inverse element [denoted by $a_v^{(*)-1}$] in the sense of the $*_v$-product. An *inner automorphism* T_v of (II.1) is given by

$$u \to T_v u = a_v *_v u *_v a_v^{(*)-1}; \qquad (a_v \in \hat{E}(N; v))$$

and has a trivial main part.

(b) *Suppose* $b_1(W) = 0$. Let A_v be an automorphism of the *Lie algebra* (II.3) of the form (VIII.1). We have for $t = 1, 2, \ldots$

$$\partial A_t(u, v) = - \sum_{r+r'=t} P(A_r u, A_{r'} v) + \sum_{2r+s=t} A_s C^{2r+1}(u, v)$$

$$- \sum_{2r+s+s'} C^{2s+1}(A_s u, A_{s'} v), \qquad (r, r' \geq 1). \qquad \text{(VIII.2)}$$

If $t = 1$, we have $\partial A_1 = 0$ and $A_1 = \mathscr{L}(Z_0)$, when $Z_0 \in L = L^*$. There is $b_0 \in N$ such that $A_1 u = P(b_0, u)$. Associate with $a_0 = \exp(-b_0/2) > 0$ the automorphisms

$$T_v^{(1)} u = a_0 *_v u *_v a_0^{(*)-1}, \qquad A_v^{(1)} = T_v^{(1)} A_v = \sum v^s A_s^{(1)}.$$

$A_v^{(1)}$ is an automorphism of the Lie algebra, with a trivial main part, such that $A_1^{(1)} = 0$ and $\partial A_2^{(1)} = 0$. There is $b_1 \in N$ such that $A_2^{(1)} u = P(b_1, u)$. Introduce for $a_1 = -b_1/2$:

$$T_v^{(2)} u = (1 + va_1) *_v u *_v (1 + va_1)^{(*)-1}, \qquad A_v^{(2)} = T_v^{(2)} A_v^{(1)} = \sum v^s A_s^{(2)},$$

where $A_1^{(2)} = A_2^{(2)} = 0$. If we proceed by recursion, we see that there is

$$c_v = \left(\prod_{s=1}^{\infty} {}_{\overline{*}} (1 + v^s a_s) \right) *_v a_0 \in \hat{E}(N; v),$$

such that $A_v u = c_v^{(*)-1} *_v u *_v c_r$, and A_v is necessarily an *inner automorphism* of the associative algebra (II.1).

(c) If $b_1(W) \neq 0$, we introduce the universal covering of W and we obtain:

Theorem   The group $\mathrm{aut}_t(*_v)$ of the automorphisms with a trivial main part of an associative algebra satisfying the assumptions (H) coincides with the group of the automorphisms of the corresponding Lie algebra which have the form

$$A_v = \mathrm{Id} + \sum_{s=1}^{\infty} v^s A_s,$$

where the A_s are differential operators that are null on the constants. If $b_1(W) = 0$, $\mathrm{aut}_t(*_v)$ coincides with the group $\mathrm{aut}_i(*_v)$ of the inner automorphisms of the associative algebra.

IX. The Main Theorem

(a) Consider now the automorphisms of the Lie algebra (VI.1), which have the form (VII.1). If $b_1(W) = 0$, we can prove that

$$A_v = T_v A_v^{(p)}, \qquad \text{(IX.1)}$$

where T_v is an inner automorphism of the associative algebra (II.1) and where $A_v^{(p)}$ is an automorphism of the Lie algebra that is even in v. It follows from the proposition of Section VII,b that $A_v^{(p)}$ is an automorphism of the associative algebra. If $b_1(W) \neq 0$, we introduce the universal covering of W and we obtain:

Main theorem   The group $aut(*_v)$ *of the automorphisms of an associative algebra satisfying the assumptions* (H) *coincides with the group of the automorphisms of the corresponding Lie algebra that have the form*

$$A_v = \left(\text{Id} + \sum_{s=1}^{\infty} v^s B_s \right) \sigma^*,$$

where σ *is a symplectomorphism and the* B_s *are differential operators null on the constants.*

(b)   The group $aut(*_v)/aut_t(*_v)$ is thus isomorphic to a subgroup H of the group $symp(W, F)$. It follows from Section V,b that

$$symp_c(W, F) \subset H. \tag{IX.2}$$

(c)   Consider in particular a $*_v$-product giving a Lie algebra *tangent to a Vey Lie algebra*

$$[u, v]_{v^2} = P(u, v) + (v^2/3!)Q^3(u, v) + \cdots.$$

The study of the automorphisms independent of v of the associative algebra shows that the symplectic connection Γ defined by Q^3 is necessarily invariant under the symplectomorphism σ of (VII.1). *The group of all these automorphisms independent of* v *is necessarily finite dimensional.*

References

[1]   Avez, A., and Lichnerowicz, A., *C. R. Acad. Sci., Ser. A* **275**, 113 (1972) Avez, A., Lichnerowicz, A., and Diaz-Miranda, A., *J. Differ. Geom.* **9**, 1 (1974).

[2]   Bayen, F., Flato, M., Fronsdal, C., Lichnerowicz, A., and Sternheimer, D., *Ann. Phys. (N.Y.)* **111**, 61 (1978).

[3]   Chevalley, C., and Eilenberg, S., *Trans. Am. Math. Soc.* **63**, 85 (1948).

[4]   Flato, M., Lichnerowicz, A., and Sternheimer, D., *Compos. Math.* **31**, 41 (1975).

[5]   Gerstenhaber, M., *Ann. Math.* **79**, 59 (1964).

[6]   Lichnerowicz, A., "Sur les algèbres formelles associées par déformation à une variété symplectique." *Ann. Mat. Pura Appl.*, to appear.

[7]   Moyal, J. E., *Proc. Cambridge Philos. Soc.* **45**, 99 (1949).

[8]   Vey, J., *Comment. Math. Helv.* **50**, 421 (1975).

[9]   Wigner, E. P., *Phys. Rev.* **40**, 749 (1932).

15

A Remark on Time-Independent Axisymmetric Fields

A. Papapetrou

Laboratoire de Physique Théorique
Institut Henri Poincaré
Paris, France

The static axisymmetric Einstein–Maxwell fields obey in the case of electrovacuum the following basic equations:

$$f\left(\nabla^2 f + \frac{1}{\rho} f_{,\rho}\right) = [ff] + 2f[VV],$$

$$f\left(\nabla^2 V + \frac{1}{\rho} V_{,\rho}\right) = [fV]. \tag{1}$$

V is the electrostatic potential and $f = g_{00}$, the metric being

$$ds^2 = -e^\gamma(d\rho^2 + dz^2) - \frac{1}{f}\rho^2 d\phi^2 + f\,dt^2.$$

Moreover,

$$\nabla^2(\ ) \equiv (\ )_{,\rho\rho} + (\ )_{,zz}; \quad [XY] \equiv X_{,\rho}Y_{,\rho} + X_{,z}Y_{,z}.$$

The stationary axisymmetric (vacuum) gravitational field obeys the equations

$$f^*\left(\nabla^2 f^* + \frac{1}{\rho} f^*_{,\rho}\right) = [f^*f^*] - \frac{1}{\rho^2}(f^*)^4[ww],$$

$$f^*\left(\nabla^2 w - \frac{1}{\rho} w_{,\rho}\right) = -2[f^*w], \tag{2}$$

217

the metric now being

$$ds^2 = -e^{\gamma^*}(d\rho^2 + dz^2) - \frac{1}{f^*}\rho^2\,d\phi^2 + f^*(dt - w\,d\phi)^2.$$

Bonnor [1] has shown that there is a one-to-one correspondence between solutions of Eqs. (1) and (2). Indeed, if we introduce in (1) new functions F and Ω,

$$f = F^2; \qquad V_{,\rho} = (if/\rho)\Omega_{,z}, \qquad V_{,z} = (-if/\rho)\Omega_{,\rho},$$

we find that F and Ω obey exactly the same equations as the functions f^* and w. Hence the theorem of Bonnor: If (f, V) is a solution of (1), then there is a corresponding solution (f^*, w) of (2) determined by the relations

$$f^* = \sqrt{f}; \quad \omega_{,\rho} = (i\rho/f)V_{,z}, \quad \omega_{,z} = (-i\rho/f)V_{,\rho}. \tag{2a}$$

Note that a real solution (f^*, w) of (2) corresponds to a solution of (1) having a real (and positive) f and imaginary V. Inversely, a real solution (f, V) of (1) corresponds to a solution of (2) having real f^* and imaginary w.

There is a second correspondence between solutions of (1) and (2) that is obtained in the following way. We introduce in (1) a new function $\tilde{F}$ by the relation

$$f = \rho^2/(\tilde{F})^2.$$

We then obtain exactly Eqs. (2) with f^* replaced by $\tilde{F}$ and w by V. Therefore, we have the following new correspondence of solutions (f, V) and (f^*, w):

$$f^* = \rho/\sqrt{f}, \qquad w = V. \tag{2b}$$

Note that now f^* and w are real when f and V are real (and $f > 0$). However we cannot now have f and $f^* \to 1$ for $r \to \infty$: Only one of the solutions (f, V) and (f^*, w) could satisfy the condition at infinity.

For an application let us consider the Weyl solution [2] of Eqs. (1) based on the assumption $f = f(V)$. One finds first

$$f = (V - \alpha)(V - \beta), \tag{3}$$

α being an arbitrary constant and $\beta = 1/\alpha$, and then finally,

$$V = \sinh(\lambda\chi)\{\lambda\cosh(\lambda\chi) + \mu\sinh(\lambda\chi)\}^{-1},$$
$$f = \lambda^2\{\lambda\cosh(\lambda\chi) + \mu\sinh(\lambda\chi)\}^{-2}, \tag{4}$$

with $\lambda = \frac{1}{2}(\alpha - \beta)$, $\mu = \frac{1}{2}(\alpha + \beta)$, and χ an arbitrary harmonic function.

The correspondence (2a) allows to determine a real solution (f^*, w) of Eqs. (2) by taking imaginary constants α, β, and an imaginary function χ.

The final result is known (see, e.g. [3]) to be the Papapetrou solution [4], which can be written in the form

$$f^* = (K \cosh \psi)^{-1}, \quad w_{,\rho} = \pm K\rho\psi_{,z}, \quad w_{,z} = \mp K\rho\psi_{,\rho}, \tag{5}$$

the constant K and the harmonic function ψ being real.

Now the correspondence (2b) determines from the Weyl solution (4) of the Einstein–Maxwell Eqs. (1) a second solution (f'^*, w') of the gravitational Eqs. (2):

$$f'^* = (\rho/\lambda)\{\lambda \cosh(\lambda\chi) + \mu \sinh(\lambda\chi)\},$$
$$w' = \sinh(\lambda\chi)\{\lambda \cosh(\lambda\chi) + \mu \sinh(\lambda\chi)\}^{-1}. \tag{6}$$

The general solution (6) is equivalent to the solution obtained by Lewis [5]. The field (f'^*, w') is real first when the constants λ, μ, and the function χ are real. It is again real in the following two cases:

(i) λ is imaginary, μ and χ real. In this case f'^* and w' are expressed in terms of trigonometric functions of a real harmonic function ψ:

$$f'^* = C\rho \cos \psi, \quad w' = \pm C^{-1} \operatorname{tg} \psi; \quad C = \text{const.} \tag{7}$$

(ii)   taking the limit $\lambda \to 0$ of (6), one finds the following solution of (2):

$$f'^* = \rho\psi, \quad w' = \pm \psi^{-1} + C. \tag{8}$$

The detailed forms (6) to (8) have been given by van Stockum [6].

One can now proceed inversely and apply the correspondence (2b) on the gravitational field (5), which has been obtained from the Weyl solution (4) by the correspondence (2a). The result is another solution (f', V') of the Eqs. (1):

$$f' = \rho^2(K \cosh \psi)^2; \quad V'_{,\rho} = \pm K\rho\psi_{,z}; \quad V'_{,z} = \mp K\rho\psi_{,\rho}. \tag{9}$$

Similarly, one can apply the correspondence (2a) on the gravitational field (6) obtained from the Weyl solution (4) by the correspondence (2b) and determine an Einstein–Maxwell field (f'', V''). However it is easy to show generally, starting from any (f, V), that $f'' = f'$ and $V'' = V'$. Indeed, starting from (f, V) we obtain by the correspondence (2a):

$$f^* = \sqrt{f}; \quad w_{,\rho} = (i\rho/f)V_{,z}; \quad w_{,z} = (-i\rho/f)V_{,\rho}, \tag{10}$$

and by the correspondence (2b):

$$f'^* = \rho/\sqrt{f}; \quad w' = V. \tag{11}$$

Now from (f^*, w) we obtain by the correspondence (2b):

$$f' = \rho^2/(f^*)^2 = \rho^2/f; \quad V' = w, \tag{12}$$

and then from (f'^*, w') by the correspondence (2a):

$$f'' = (f'^*)^2 = \rho^2/f = f'; \tag{13a}$$

$$w'_{,\rho} = (i\rho/f'')V''_{,z}; \qquad w'_z = (-i\rho/f'')V''_{,\rho}. \tag{13b}$$

Because of (11), (13), and (10), Eqs. (13) reduce to

$$V''_{,z} = w_{,z}, \qquad V''_{,\rho} = w_{,\rho} \leftrightarrow V'' = w = V' \tag{14}$$

and consequently $(f'', V'') \equiv (f', V')$.

It should be noted that the result expressed by the relation (2b) can also be obtained in a different way, e.g., by combining the transformation obtained by Kramer and Neugebauer [7] for stationary vacuum fields with the correspondence (2a) of Bonnor.† The result obtained in this paper is that starting from any solution, e.g., a solution (f, V) of Eqs. (1), we arrive at a pair of solutions (f, V) and (f', V') of Eqs. (1) as well as solutions (f^*, w) and (f'^*, w') of Eqs. (2) corresponding, respectively, by means of the relations (2a) and (2b). Note that the functions V and V' as well as w and w' are orthogonal:

$$[VV'] = 0 = [ww'], \tag{15}$$

while the functions f, f' and f^*, f'^* satisfy the relations

$$ff' = \rho^2, \qquad f^*f'^* = \rho. \tag{16}$$

References

[1] Bonnor, W., Z. Phys. **190**, 444 (1966).
[2] Weyl, H., Ann. Phys. (Leipzig) **54**, 117 (1917).
[3] Kinnersley, W., Proc. GR7, Tel.-Aviv University, 1974.
[4] Papapetrou, A., Ann. Phys. (Leipzig) **12**, 309 (1953).
[5] Lewis, T., Proc. R. Soc. London, Ser. A **136**, 176 (1932).
[6] van Stockum, W. J., Proc. R. Soc. Edinburgh **57**, 135 (1937).
[7] Kramer, D., and Neugebauer, G., Commun. Math. Phys. **10**, 132 (1968).

† I thank Dr. E. Herlt for this remark.

16

Values and Arguments in Homogeneous Spaces†

Charles W. Misner

Department of Physics and Astronomy
University of Maryland
College Park, Maryland

Abstract

Curved homogeneous spaces have played important roles in relativity and cosmology. A number of well-known applications are listed with Taub's contributions noted. In these familiar cases the homogeneous space was space–time or a slice of it, and served as the domain (or argument manifold) for the physical functions or fields. Another use of homogeneous spaces is then proposed, using them as the range (or manifold of values) of fields defined on flat (or curved) space–time. Few physical examples are yet known of physical fields taking values in a curved manifold, but the mathematics of harmonic maps, their aptitude for expressing symmetry breaking, and suggestive relationships to gauge theories make exploration of this idea attractive. (A brief introduction to harmonic maps is given, summarizing Misner [16] and Misner, Schild Memorial Lectures, 1978.) The style or technique of symmetry breaking that harmonic maps implement is a style that is implicit, but not often recognized, in general relativity. The main new result of this paper is a simple proof, by analogy to the geodesic equations, that if a field ϕ satisfying a harmonic mapping field equation takes its values in a homogeneous space, then for every Killing vector ξ^A in that space, the quantity

$$J_\mu(x) = \xi_A[\phi(x)](\partial \phi^A / \partial x^\mu)$$

is a conserved current $J^\mu_{;\mu} = 0$ even if space–time itself has no Killing vectors. (A third section of the lecture concerning harmonic connections is omitted from the text, as it is given briefly in Misner [16] and more fully in a paper in draft.)

Abe Taub is a great friend and teacher. He taught me to appreciate the technique of using Killing constants of motion in the geodesic equations that we used in our collaboration [1] and that I apply here to a different

† Supported in part by National Science Foundation Grant PHY78-09658.

problem. I do not know exactly what he taught other collaborators technically, but I have seen in the work of his postdoctoral associates the effects of an experienced and gifted mentor at work. In addition to his personal influence on many collaborators and associates, Abe has also taught widely by his insightful and innovative researches to which his colleagues return year after year for renewed edification.

I. Introduction

Figure 1 illustrates the title of this contribution. Homogeneous spaces are most familiar in physics as the domain spaces for real or vector-valued functions $f(x)$ whose arguments x lie in the homogeneous space. This is the case, for example, in the spatially homogeneous cosmologies that Taub has studied [2]. But I will also describe speculative field theories where the spacetime point x may be in flat Minkowski space, but fields $\phi(x)$ are considered where the field value is a point in a nonflat homogeneous space. For example, ϕ could be a direction in space, i.e., a point on the two-sphere S^2. Or the field value ϕ could be a preferred (symmetry breaking) direction in some internal symmetry space; e.g., ϕ could be a point in the homogeneous space $SU(3)/S(U_2 \times U_1)$—which is the complex projective plane—in a theory where $SU(3)$ symmetry is seen broken down to $SU(2) \times U(1)$.

Before proceeding further, let us recall some definitions. A *homogeneous space* consists of a differentiable manifold M together with a group G of

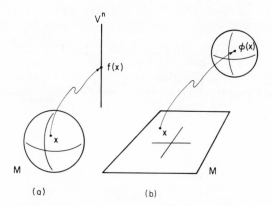

Fig. 1 Nonflat homogeneous spaces, here symbolized by a sphere, occur most familiarly as the space sections in homogeneous cosmologies. Then, for a field $f(x)$ as in (a), it is the argument x that lies in the homogeneous space. In theories of harmonic maps, however, the nonflat homogeneous space occurs (as in b) as the set of values ϕ of fields that may be defined over flat Minkowski space.

transformations under which all points of M are geometrically equivalent. More precisely, one supposes that each $s \in G$ defines a (differentiable) mapping $s: M \to M: x \to sx$ of M onto itself with various properties. One wants the mapping $G \times M \to M: (s, x) \to sx$ to be differentiable and satisfy $s(tx) = (st)x$. Then $t = s^{-1}$ (group inverse) supplies an inverse mapping to s so that all transformations are diffeomorphisms. In the applications considered here the underlying manifold M is also a Riemannian manifold and the mappings s are isometries as well. To focus the discussion on the transformations $M \to M$ rather than unrelated group structure, one requires that the group G act *effectively*, i.e., that if $sx = x$ for all $x \in M$ then s is the group identity element. Finally one requires that all points in M be equivalent under G by insisting that for some (hence any) $x \in M$, the set of its translates be M, i.e., one requires that $Gx \equiv \{sx \,|\, s \in G\}$ be M for $x \in M$. Then one finds that the collections of spaces M that can be homogeneous under a group G is catalogued by the closed subgroups of G, and each such homogeneous space is isomorphic to G/H for some subgroup H. The required subgroup can be found as the isotropy subgroup at a point: For any $x \in M$ the isotropy subgroup H_x is $\{t \in G \,|\, tx = x\}$ and all such H_x are isomorphic—$H_{sx} = sH_x s^{-1}$. One then finds that M is diffeomorphic to G/H_x which is a coset space

$$G/H = \{sH \subset G \,|\, s \in G\},$$

where the cosets $sH = \{st \,|\, t \in H\}$ are equivalence classes under the subgroup H. A simple example is

$$P^2 = SO(3)/SO(2).$$

Here a rotation from $SO(3)$ can be identified by three Euler angles $\psi\theta\phi$ and is a rotation by ψ about an axis in the direction $\theta\phi$. By factoring out the $SO(2)$ subgroup of rotations about a fixed axis, one arrives at the projective space P^2 as the manifold of possible axes. (The coset space G/H does not inherit a group structure from G unless H is a normal subgroup of G.)

II.   Homogeneous Cosmologies

Homogeneous spaces are familiar to relativists from studies of homogeneous cosmologies. In the original Einstein universe the space–time manifold was $S^3 \times R^1$, which is homogeneous under the action of the group $SO(4) \times E^1$ where E^1 is the additive group of real numbers (time translations). Of course Minkowski space is also a (flat) homogeneous space under the action of the Poincare group. Other four-dimensionally homogeneous cosmologies include the steady-state universe, the deSitter universe, and the

Gödel universe. They have been catalogued [3] and now illuminate our physical understanding of cosmology by providing examples of conceptual alternatives that Nature apparently chose not to implement. Other cosmological models have only a three-dimensional homogeneity. There the transformation group G does not make space–time M a homogeneous space, but each event $x \in M$ is contained in a three-dimensional submanifold $Gx \subset M$ that is homogeneous (and usually spacelike). These spacially homogeneous cosmologies include the standard Friedmann–Robertson–Walker cosmologies where each point has an isotropy group $H = O(3)$, the Taub anisotropic cosmological models where generically $H = \{e\}$ is the trivial isotropy group, and some other models such as the Kantowski–Sachs [4] universe where the spacelike surfaces are $S^2 \times R = SU(2) \times E^1/U(1)$.

The Taub anisotropic cosmological models use space–time manifolds $M = G \times R$ where the hypersurfaces $G \times \{t\}$ are isomorphic to some three-dimensional group manifold G. These three-dimensional Lie groups had been classified by Bianchi; Taub pointed out their suitability as cosmological models and developed the technical apparatus for making use of them [2]. These Taub cosmological models have played important roles in the development of theoretical cosmology. Their importance derives from two properties. First, they are alternatives to and generalizations of the standard Friedmann models. Second, they are eminently computable. Consequently they have provided a workshop where cosmologists could sharpen their intuitions, test hypotheses, and try out new ideas or, from examples, have new ideas suggested.

These Taub model universes became particularly important with the discovery of the cosmic microwave background radiation. The measured isotropy of this radiation was so precisely zero that it demanded more than a hypothesis of simplicity as an explanation, and Taub's set of anisotropic but still homogeneous cosmological models became the natural arena for this study [5]. It has led to a deep mystery, the horizon paradox, that appears as portentious of future discoveries as the Olbers' paradox should have been a century ago. The horizon paradox arises from the observation that widely separated regions of the sky (many degrees apart) show the same microwave background temperature. Yet neither in the Friedmann standard models [6], nor in Taub's generalizations [7], can the distinct masses of primordial plasma emitting this observed radiation have known of each other. The time available in these models is simply too short, from the big bang initial singularity to the epoch where the primordial "fireball" becomes transparent and emits the radiation seen now, for any causality limited interaction to have regulated the temperature differences between different presently observable regions. Without the studies based on Taub's models, the inadequacy of our understanding of the microwave isotropy might have been dismissed as an unsurprising consequence of using the simple Friedmann models. Instead, one now worries that a completely new picture of the first microsecond of the big bang will be required, and that modifications to Einstein's equations will be needed, or at least their quantization.

The paradoxical isotropy of the cosmic background radiation is just one of several questions that together lead to the *great cosmological question* which is "Can the beginnings of the Universe, i.e., the choice of initial conditions, be made the subject of scientific discourse, rather than merely the object of historical discovery?" Traditional science has treated either theories of equilibrium (thermodynamics, electrostatics, etc.) or of evolution (Newton, Maxwell, etc.) but relativistic cosmology now seems to find both these approaches inadequate. The Taub cosmological models have been involved in other work bearing on this question in addition to studies of the horizon paradox. What I call the "steady-state big bang" model [8] is precisely a way of discussing the initial conditions of the Universe that was suggested by the dynamics of some Taub cosmological models. Another approach to discussions of initial conditions, the anthropic principle, has also used examples (Wheeler, Chapter 5, this volume; Shepley, Chapter 6, this volume) built from the Taub models. The strongest pressure behind the great cosmological question comes from the singularity theorems [9] that make something like the big bang unavoidable in current theories. As these theorems were being developed, the Taub models provided counterexamples that were helpful in shaping the statements to be proven. Other outgrowths of Taub's explorations of spacially homogeneous cosmologies include the development of Hamiltonian techniques for studying cosmological models [10], and the quantum cosmology explorations of simplified models [11] of quantized gravitational fields.

III. Harmonic Maps

Now consider homogeneous spaces as the range of a function rather than its domain, as in the second sketch in Fig. 1. The symmetry group G then has nothing to do with the symmetry of spacetime. (I shall assume for simplicity here that space–time is Minkowski space, although it could equally well be an inhomogeneous cosmology without any symmetry.) Most physical fields are vectors or tensors (including scalars and spinors) and take their values in a vector space. Because linear relationships are defined in vector spaces, it is always possible to write a linear field equation for the field, and these linear equations are usually seen as the starting point of the theory, with nonlinear interactions then added. General relativity is the one familiar exception. Two metrics can be combined as tensors $ag_{\mu\nu} + bh_{\mu\nu}$ but the result is not necessarily a metric—it may be degenerate or have the wrong signature. Corresponding to this lack of linear operations on the space of (Minkowski signature) metrics is the subtlety of using a linear wave equation as a starting point in formulating general relativity—a good such derivation came only after half a century [12]. Harmonic maps are a general class of wave equations for fields where there is no natural concept of "linear," but where nonlinear field equations with many similarities to the Einstein equation can be readily formulated. I am advertising harmonic maps as a field of mathematics in search of applications. A substantial part of my motivation for expecting they may play some role in physics is merely a Pythagorean prejudice, a feeling that a beautiful piece of mathematics and geometry, for which many limiting cases are important in physics, should not have been ignored by Nature.

Harmonic maps were defined and surveyed first by Fuller [13]. I have read also the early work by Eells and Sampson [14], and there is a recent extensive review [15]. A harmonic map is a differentiable mapping

$$\phi: M \to M': x \to \phi(x) \tag{1}$$

from one Riemannian manifold M to another one M' satisfying a particular field equation. For present purposes we think of M as (Minkowski) space–time, and of M' as some homogeneous space with a group invariant metric

$$dL^2 = G_{AB}(\phi)d\phi^A \, d\phi^B. \tag{2}$$

The harmonic mapping field equations are derived from a variation principle using an action functional that is just $\int \|d\phi\|^2 *1$ where $d\phi$ is the differential of the mapping ϕ. Using local coordinates x^μ on M and ϕ^A on M', this action integral is

$$I = -\frac{c^3}{2\gamma^2} \int G_{AB}(\phi) \frac{\partial \phi^A}{\partial x^\mu} \frac{\partial \phi^B}{\partial x^\nu} g^{\mu\nu}(x)\sqrt{|g|}d^4x, \tag{3}$$

where $g_{\mu\nu}$ can be the Minkowski metric on M, but G_{AB} should be curved to give the interesting applications.

As an introduction for physicists has now been published [16], the description of harmonic maps here is very brief, with bare mention of examples from the Einstein equations and the nonlinear σ model. Harmonic functions, linear wave equations, and the geodesic equation are all special cases of harmonic maps. In general one may think of harmonic maps as governed by field equations that have the light-cone properties of linear wave equations, but nonlinearities similar to geodesic equations. Two examples may be helpful. For the first, let M' be the two-sphere S^2 with the familiar $\theta\phi$ spherical coordinates capitalized to remind us that they are field values, not space–time coordinates. The action integral and field equations for these two fields Θ and Φ are then

$$I = \tfrac{1}{2} \int [(\nabla\Theta)^2 + \sin^2 \Theta(\nabla\Phi)^2] d^4 x,$$

$$\Delta\Theta + \sin \Theta \cos \Theta(\nabla\Phi)^2 = 0, \tag{4}$$

$$\Delta\Phi - 2 \cot \Theta(\nabla\Theta) \cdot (\nabla\Phi) = 0.$$

These field equations are closely related to the Sine–Gordon equation. Note the simple d'Alembertian $\Delta \equiv -\partial_\mu\partial^\mu$ in the leading term, and the strong nonlinearities in the other terms. A solution $\Theta(x^\mu)$, $\Phi(x^\mu)$ defines a map $\phi: M \to S^2$ from Minkowski space to S^2. For a second example we define a field that gives mappings

$$\phi: M \to SU(3)/S(U_2 \times U_1), \tag{5}$$

where $S(U_2 \times U_1)$ are the unit determinant matrices from $U(2) \times U(1)$, i.e., the matrices in $SU(3)$ that are in 2×2 plus 1×1 block diagonal form. This subgroup consists of all matrices in $SU(3)$ that commute with the projection matrix $\phi_B^A = \delta_3^A\delta_B^3$. A coset of this subgroup is just the $SU(3)$ matrices that commute with some other one-dimensional projection ϕ characterized among 3×3 matrices by

$$\phi^2 = \phi = \phi\dagger, \qquad \text{tr } \phi = 1. \tag{6}$$

The $\phi^2 = \phi$ condition is clearly not consistent with linearly combining ϕs, so a nonlinear wave equation will be required to define a ϕ field. (This ϕ field picks out dynamically a preferred axis in some three-dimensional internal symmetry space.) There is a natural metric on the manifold of such projection matrices, defined by $dL^2 = (\tfrac{1}{2})\text{tr}(d\phi\dagger \, d\phi)$ subject to the constraints in (III.6). Using this metric as $G_{AB}(\phi)$ in (3) leads to the harmonic mapping wave equation

$$\phi(\Delta\phi)(1 - \phi) = 0$$

$$\tag{7}$$

or

$$\phi_B^A(\Delta\phi_C^B)(\delta_D^C - \phi_D^C) = 0.$$

This equation will preserve the $\phi^2 = \phi$ and other constraints while propagating ϕ waves through Minkowski space–time along the light cones of the d'Alembertian Δ.

In the oral presentation some attention was paid to the role harmonic maps may have, especially in quantization, as models for the Einstein equations in view of the fact that their action integral

$$I_{HM} = -(c^3/\gamma^2) \int G(\phi)(\partial\phi/\partial x)^2 \, d^4x \qquad (8a)$$

has a strong formal resemblance to the Einstein action

$$I_{GR} = -(c^3/G) \int g^- g^- g^- (\partial g_-/\partial x)^2 \, d^4x \qquad (8b)$$

that is more striking even than the similarities of the Einstein theory to the Yang–Mills theory

$$I_{YM} = -(\hbar^2/e^2) \int (\partial A + A^2)^2 \, d^4x, \qquad (8c)$$

which has already been a very fruitful analogy. This discussion is omitted here, as also are the remarks on "Harmonic connections," i.e., the possible use of harmonic mapping equations as (inequivalent) alternates to the Yang–Mills equations as field equations defining a gauge connection. Both of these topics are treated in a now-published paper [16]. I proceed to note that conservation laws in the form of conserved currents are available for the harmonic mapping equation whenever the range space M' admits Killing vectors. They arise in complete analogy to the way in which constants of motion arise for the geodesic equation whenever space–time admits Killing vectors. Note, however, that the conserved currents we find here exist independently of the symmetries (or lack of such) that may exist in space–time, and would occur for harmonic maps into homogeneous spaces even from an inhomogeneous cosmology that had no space–time symmetries.

The close parallel of the harmonic mapping equation to the geodesic equation is seen by considering the case of harmonic maps from a one-dimensional manifold; then the harmonic mapping equation is nothing but the geodesic equation with proper distance along the curve in M' being identified with length along the line M. Let us write the harmonic mapping equation in a notation that can make use of this parallel. The equation

$$\phi_\mu^{A;\mu} = 0 \qquad (9)$$

is the harmonic mapping equation in one notation. The quantities $\phi_\mu^A = \partial\phi^A/\partial x^\mu$ are the components of $d\phi$ or also the components of $\phi_* \partial_\mu \in M'_{\phi(x)}$, the image tangent to M' of the vector ∂_μ tangent to M. The covariant derivative in Eq. (9) regards them as components of $d\phi$ and thus contains correction terms both for the motion of basis vector ∂_μ tangent to M, and for basis vectors ∂_A tangent to M'. Thus

$$
\begin{aligned}
0 &= \phi_\mu^{A;\mu} \\
&= g^{\mu\nu}(\phi_{\mu,\nu}^A - \phi_\sigma^A \Gamma_{\mu\nu}^\sigma + \phi_\mu^B \Gamma_{B\nu}^A) \\
&= g^{\mu\nu}(\phi_{\mu,\nu}^A - \phi_\sigma^A \Gamma_{\mu\nu}^\sigma + \phi_\mu^B \phi_\nu^C \Gamma_{BC}^A(\phi(x))).
\end{aligned}
\tag{10}
$$

In this last form the analogy to the geodesic equation

$$
0 = d_t^2 \phi^A + d_t \phi^B \, d_t \phi^C \Gamma_{BC}^A(\phi(t))
\tag{11}
$$

is clear. (The quantities $\Gamma_{B\nu}^A(x)$ in (10) are the components of the connection in the ϕ-induced bundle $\phi^* TM'$ over M that derive from the Riemannian connection Γ_{BC}^A in the tangent bundle TM' over M'.)

If ξ^A is a Killing vector on M', so

$$
\xi_{A;B} + \xi_{B;A} = 0,
\tag{12}
$$

then $p = \xi_A d_t \phi^A$ is a familiar constant of motion for geodesics on M'.

Theorem   Let $\phi: M \to M'$ be a harmonic map, and ξ_A a Killing vector on M', then

$$
J_\mu \equiv \xi_A \phi_\mu^A
\tag{13}
$$

is a conserved current on M, i.e., $J^\mu_{;\mu} = 0$.

The proof of this theorem is a simple computation as in the case of the geodesic equation

$$
\begin{aligned}
J_\mu^{;\mu} &= g^{\mu\nu} J_{\mu;\nu} \\
&= g^{\mu\nu}(\xi_{A;B} \phi_\mu^A \phi_\nu^B + \xi_A \phi_{\mu;\nu}^A) \\
&= 0.
\end{aligned}
\tag{14}
$$

The last term here vanishes by the harmonic mapping equation, while the first term involving $\xi_{A;B}$ vanishes by the Killing equation and the symmetry of $g^{\mu\nu}\phi_\mu^A \phi_\nu^B$ in A and B. ∎

Broken symmetry is a familiar theme in elementary particle theory, but is not often recognized in general relativity. Consideration of harmonic maps leads one to recognize general relativity as a theory of broken symmetry. The larger symmetry with which the theory begins is general covariance seen as the local symmetry of the tangent frame bundle over space–time under the action of the $GL(4)$ group: $\{e_a(x)\} \to \{s_a^b e_b(x)\}$. The reduced symmetry that is shown by gravitational systems is local Lorentz invariance where certain tangent frames $\{e_m(x)\} = \{e_0, e_1, e_2, e_3\}$ are distinguished as "orthonormal" and among them only Lorentz group $O(3, 1)$ transformations preserve the structure of the theory. To pick the class of orthonormal frames is to pick a coset of $O(3, 1)$ in $GL(4)$: For each such choice all orthonormal frames are of the form $\{\Lambda^a_m s^b_a e_b(x)\}$ with e_b and $s \in GL(4)$ fixed while Λ varies within $O(3, 1)$. Achieving a reduction of the local symmetry group from $GL(4)$ to $O(3, 1)$ is thus achieved by including in our theory a field g whose values are classes of orthonormal frames, i.e., points in $GL(4)/O(3, 1)$. [The usual $g_{\mu\nu}$ metric components are coordinates on this space and are defined at any frame class $\{e_a\} \equiv \{\Lambda^b_a e_b\}$ by $g^{\mu\nu} = \langle dx^\mu, e_a \rangle \langle dx^\nu, e_b \rangle \eta^{ab}$ with η^{ab} the Minkowski matrix diag$(-1, 1, 1, 1)$.] Thus general relativity is a theory of maps of a (metricless) space–time manifold M into the curved homogeneous manifold $GL(4)/O(3, 1)$, or more precisely a theory of sections of a bundle with this fiber. The lack of an *a priori* connection in this bundle prevents one's using a harmonic mapping equation as a field equation for g in place of the Einstein equation. But it is clear that there is a strong correspondence between the structure of the Einstein theory and of harmonic maps into range spaces G/H for other pairs (G, H) of group/subgroup representing an underlying large symmetry G broken to a more strongly realized smaller symmetry H. If we call the introduction of a G/H-valued field the "Einstein style of symmetry breaking," then a central question for the theory of harmonic maps is whether this Einstein style symmetry breaking is employed by Nature elsewhere than in gravitation. Examples [16] show that Einstein-style symmetry breaking can promote *a priori* massless Yang–Mills connections A_μ to massive vector fields W_μ while breaking symmetries, all in a fashion similar to the usual Higgs mechanism in gauge theories. The Einstein method has the advantage of leaving no scalar Higgs mesons to be found, but the disadvantage of presenting obstacles to unambiguous quantization comparable to the obstacles found in general relativity.

The rich structure of harmonic mapping theories—including both topological conservation laws [16] and dynamic one (such as the conserved currents discussed above), and close parallels to both general relativity and particle theory concepts—suggests that further study of these theories is in order.

References

[1]   Misner, C. W., and Taub, A. H., *Sov. Phys.—JETP* **28**, 122 (1969).
[2]   Taub, A. H., *Ann. Math.* **53**, 472 (1951).
[3]   Ryan, M. P., Jr., and Shepley, L. C., "Homogeneous Relativistic Cosmologies." Princeton Univ. Press, Princeton, New Jersey, 1975. (See Chapter 7 for a survey and for the original references. The Taub models are described in Chapters 6, 8, and 9).
[4]   Kantowski, R., and Sachs, R. K., *J. Math. Phys.* **7**, 443 (1967).
[5]   Misner, C. W., *Astrophys. J.* **151**, 431 (1968); see also MacCallum, M. A. H., *Proc. Cracow Int. Summer Sch. Cosmol., 1st, 1979* (1980).
[6]   Misner, C. W., *Phys. Rev. Lett.* **19**, 533 (1967).

[7] Doroshkevich, A. G., Lukash, V., and Novikov, I. D., *Zh. Eksp. Teor. Fiz.* **60**, 1201 (1971); Engl. transl. *Sov. Phys.—JETP* **33**, 649 (1971); Chitre, D., *Phys. Rev. D* **6**, 3390 (1972).

[8] Misner, C. W., Thorne, K. S., and Wheeler, J. A., "Gravitation," Box 30.1. Freeman, San Francisco, California, 1973; see also Misner, C. W., *in* "Cosmology, History, and Theology" (W. Yourgrau and A. D. Breck, eds.), pp. 87–93. Plenum, New York, 1977; Misner, C. W., *in* "Some Strangeness in the Proportion: A Centennial Symposium to Celebrate the Achievements of Albert Einstein" (H. Woolf, ed.), pp. 405–415. Addison-Wesley, Reading, Massachusetts, 1980.

[9] Hawking, S. W., and Ellis, G. F. R., "The Large Scale Structure of Space-Time." Cambridge Univ. Press, London and New York, 1973.

[10] Ryan, M., "Hamiltonian Cosmology," Lecture Notes in Physics, No. 13. Springer-Verlag, Berlin and New York, 1972.

[11] MacCallum, M. A. H., *in* "Quantam Gravity" (C. J. Isham, R. Penrose, and D. W. Sciama, eds.), pp. 174–218. Oxford Univ. Press, London and New York, 1975.

[12] Deser, S., *Gen. Relativ. Grav.* **1**, 9 (1970), which includes references to earlier work of Gupta and Feynman.

[13] Fuller, F. B., *Proc. Natl. Acad. Sci. U.S.A.* **40**, 987 (1954).

[14] Eells, J., Jr., and Sampson, J. H., *Am. J. Math.* **86**, 109 (1964).

[15] Eells, J., and Lemaire, L., *Bull. London Math. Soc.* **10**, 1 (1978).

[16] Misner, C. W., *Phys. Rev. D* **18**, 4510 (1978).

Curriculum Vitae of Abraham Haskel Taub

Born, Chicago, February 1, 1911, son of Joseph Haskel and Mary (née Sherman) Taub.

Married, Cecilia Vaslow, December 26, 1933; children, Mara, Nadine, and Haskel Joseph.

Educated, University of Chicago (B.S., 1931) and Princeton University (Ph.D., 1935).

Intructor in Mathematics, Princeton University, 1934–1935.

Assistant, Institute for Advanced Study, 1935–1936.

Instructor to full Professor, University of Washington at Seattle, 1936–1948.

Professor of Mathematics, University of Illinois at Urbana, 1948–1964.

Head of Digital Computer Laboratory, University of Illinois at Urbana, 1961–1964.

Professor of Mathematics and Director of the Computer Center, University of California at Berkeley, 1964–1967.

Professor of Mathematics, University of California at Berkeley, 1967–1978.

Member, Institute for Advanced Study, 1940–1941, 1947–1948, 1962–1963.

Visiting Professor, Collège de France, Paris, France, 1967.

Recipient President's Certificate of Merit, 1946.

Medal from the City of Lille, France, 1969.

Guggenheim Fellow, 1947–1948, 1958.

Member, American Mathematical Society, American Physical Society, and American Association for the Advancement of Science.

List of Publications by A. H. Taub

1. Projective differentiation of spinors (with O. Veblen), *Nat. Acad. Sci. Proc.* **20**, 85–92 (1934).
2. Dirac equation in projective relativity (with O. Veblen and J. von Neumann), *Nat. Acad. Sci. Proc.* **20**, 383–396 (1934).
3. Quantum equations in cosmological spaces, *Phys. Rev.* **51**, 512–525 (1937).
4. Spin representation of inversions, *Bull. Amer. Math. Soc.* **44**, 860–865 (1938).
5. Tensor equations equivalent to the Dirac equation, *Ann. of Math.* **40**, 937–947 (1939).
6. Spinor equations for the meson and their solution when no field is present, *Phys. Rev.* **56**, 799–908 (1939).
7. Solution of equations for particles of spin zero or one when no field is present, *Phys. Rev.* **57**, 807–814 (1940).
8. The acceleration of the Dirac electron, six studies in mathematics, *U. of Washington Publications in Mathematics* **2**, No. 3, 41–44 (1940).
9. Interaction of progressive refraction waves, *Ann. of Math.* **47**, 811–828 (1946).
10. Refraction of plane shock waves, *Phys. Rev.* **72**, 51–60 (1947).
11. Some numerical results on refraction of plane shock waves, Princeton University Report on Contract Nori-105 Task II (1947).
12. On Hamilton's principle for perfect compressible fluids, *Proc. of First Symposium of Appl. Math. Am. Math. Soc.* (1949).
13. Orbits of charged particles in constant fields, *Phys. Rev.* **73**, 786–798 (1948).
14. Characterization of conformally flat spaces, *Bull. Amer. Math. Soc.* **55**, 85–89 (1949).
15. Relativistic Rankine-Hugoniot equations, *Phys. Rev.* **74**, 328–334 (1948).
16. A special method for solving the Dirac equations, *Rev. Modern Phys.* **21**, 388–392 (1949).
17. Interaction of shock waves (with W. Bleakney) *Rev. Modern Phys.* **21**, 584–605 (1949).
18. Theory of the parallel plane diode, *J. Appl. Phys.* **21**, 974–980 (1950).
19. Empty space-times admitting a three parameter group of motions, *Ann. of Math.* **53**, 472–490 (1951).
19a. Empty space-times admitting a three parameter group of motions, *Proceedings of the International Congress of Mathematics,* 1950.
20. A sampling method for solving the equations of compressible flow in a permeable medium, *Proc. of First Conf. on Fluid Dynamics,* Ann Arbor, Mich. (1950).
21. The Mach reflection of shock waves at nearly glancing incidence (with C. H. Fletcher & W. Bleakney) *Rev. Modern Phys.* **23**, 271–286 (1951).
22. Curved shocks in pseudo-stationary flows, *Ann. of Math.* **58**, 501–527 (1953).
23. Numerical results on the shock configuration in math reflection (with R. D. Clutterham), *Proc. of Sixth Symposium in Appl. Math., Am. Math. Soc.,* McGraw Hill, New York (1956) pp. 45–58.
24. A general relativistic variational principle for perfect fluids, *Phys. Rev.* **94**, 1468–1470 (1954).
25. Singularities on shocks, *Amer. Math. Monthly* **61**, 11–12 (1954).
26. Born-type rigid motion in relativity (with G. Salzman) *Phys. Rev.* **95**, 1659–1669 (1954).
27. Determination of flows behind stationary and pseudo-stationary shocks, *Ann. of Math.* **62**, 300–325 (1955).

28. Automata in production and management, *Illinois Business Review.* **13**, 8–9 (1956).
29. Isentropic hydrodynamics in plane symmetric space time, *Phys. Rev.* **103**, 454–467 (1956).
30. Approximate solutions of the Einstein equations for isentropic motions of plane-symmetric distributions of perfect fluids, *Phys. Rev.* **107**, 844–900 (1957).
31. Numerical solutions of Sturm-Liouville differential equations (with C. C. Farrington, R. T. Gregory) MTAC XI, No. 59, 131–150 (1957).
32. Singular hypersurfaces in general relativity, *IJM* **1**, No. 3, 370–388 (1957).
33. Wave propogation in fluids, Chapter 4 Part III, "Handbook of Physics" (E. U. Condon and Hugh Odishau, eds.). McGraw-Hill, New York, 1958.
34. On circulation in relativistic hydrodynamics, Archive for Rational Mechanics and Analysis, **3**, 312–324 (1959).
35. A study of a numerical solution to a two-dimensional hydrodynamical problem (with A. Blair, N. Metropolis, J. von Neumann, M. Tsingou). "Mathematical Tables and Other Aids to Computation," Vol. XIII, No. 67, July, 1959, pp. 145–184.
36. Approximate stress energy tensor for gravitational fields, *J. Math. Phys.* **2**, 787–793 (1961).
37. On spherically symmetric distributions of incompressible fluids. In "Recent Developments in General Relativity" (Pergamon Press, New York, 1962).
38. Small motions of a spherically symmetric distribution of matter; Les theories relativistes de la gravitation (Colloques Interaction aux Royaumont (959), Paris 1962, pp. 173–191 (1962).
39. On Thomas' result concerning the geodesic hypothesis, *Proc. Nat. Acad. Sci.* **48**, 150–151 (1962).
40. H. P. Robertson (1903–1961), *J. Soc. Indust. App. Math.* **4**, 739–750 (1962).
40a. H. P. Robertson (1903–1961), *Nature* **192**, 797–798 (1961).
41. Hydrodynamics and general relativity. In "Relativistic Fluid Mechanics and Magneto-hydrodynamics." Academic Press, New York, pp. 21–28 (1963).
41a. Hydrodynamics and general relativity, Sandia Corporation, Research Colloquium, March 1959.
42. Motion of test bodies in general relativity, *J. Math. Phys.* **5**, 112–119 (1964).
43. The motion of multipoles in general relativity. Atti del Convegno Sulla Relatività Generale: Problemi dell'Energiea e Onde Gravitazionale. (IV Contenario della Nascita di Galileo Galilei) Florence 9–12 Sept. 1964.
44. Relativistic gas spheres at constant entropy. In *Proceedings of the Second Texas Symposium on Relativistic Astrophysics.* University of Chicago Press, Chicago, 1965.
45. The Riemann–Christoffel tensor and tetrad and self-dual formalisms. In "Perspectives in Geometry and Relativity," B. Hoffman, ed. University of Indiana Press, Bloomington, 1965, pp. 360–368.
46. Equations of motion of text particles, *Proceedings of the International Conference on General Relativity and Gravitation.* London (1965).
47. Relativistic hydrodynamics. Lectures in Applied Mathematics Vol. 8 and Battelle Rencontres, Relativity Theory and Astrophysics and Relativity and Cosmology, pp. 170–193 (1967).
48. On time-sharing systems, design and use, *Proceedings of the IBM Scientific Computing Symposium on Man-Machine Communication,* pp. 9–16 (May 1965).
49. Space-times containing perfect fluids and having a vanishing conformal divergence, *Comm. Math. Phys.* **5**, 237–256 (1968).
50. A singularity-free empty universe (with C. W. Misner), *J.E.T.P.* **55**, 233–254 (1968).
51. "Computer Science Lectures in Applied Math," vol. 12 of Mathematics of Decision Sciences Part 2, 371–381 (1968).

52. Restricted motions of gravitating spheres, *Ann. Inst. H. Poincaré*, **IX**, 153–179 (1968).
53. Stability of Fluid Motions and Variational Principles, Fluides et Champ Gravitational En Relativité Générale Colloques Inter. du Centre National de Recherche Scientifique pp. 57–71, Paris 1969; also in Hyperbolic Eqs. and Waves, Battelle Seattle 1968 Rencontres, Springer-Verlag, Berlin and New York, 1970.
54. Stability of General Relativistic Gaseous Masses and Variational Principles, *Comm. Math. Phys.* **15**, 235–254 (1969).
55. "General Relativistic Hydrodynamics, Hyperbolic Equations, and Waves," Battelle Seattle 1968 Rencontres, Springer-Verlag, Berlin and New York, pp. 22–23 (1970).
56. Variational principles and relativistic magneto-hydrodynamics, La magnetohydrodynamique classique et relativiste, Colloque Internationaux du Centre National de la Recherche Scientifique No. 184 Lille 16–20 Juin 1969, Paris 1970, pp. 189–200.
57. Spherically symmetric similarity solutions of the Einstein field equations for a perfect fluid, *Comm. Math. Phys.* **21**, 1–40 (1971).
58. Variational principles in general relativity; relativistic fluid dynamics, Centro Internazionale Matematico Estivo, Bressanone 7–16 July 1970, Edizioni Cremonese, Roma 1971.
59. Relativistic hydrodynamics, in Studies in Applied Mathematics; Mathematical Association of American Studies in Mathematics Vol. 7 (1971), (A. H. Taub, editor).
60. Variational principles and spatially-homogeneous universes, including rotation (with M. A. H. MacCallum) *Comm. Math. Phys.* **25**, 173–189 (1972).
61. Plane-symmetric similarity solutions for self-gravitating fluids. In "General Relativity" (papers in honor of J. L. Synge; L. O'Raefeartaigh, ed.) Clarendon Press, Oxford (1972), pp. 133–150.
62. Plane symmetric self-gravitating fluids with pressure equal to energy density (with Romualdo Tabensky) *Comm. Math. Phys.* **29**, 61–77 (1973).
63. General relativistic shock waves in fluids for which pressure equals energy density, *Comm. Math. Phys.* **29**, 79–88 (1973).
64. The averaged Lagrangian and high-frequency gravitational wave (with M. A. H. MacCallum), *Comm. Math. Phys.* **30**, 153–169 (1973).
65. On Variational Principles in General Relativity in Gravitation (papers in honor of A. Z. Petrov). Publishing House "Naukova Dumka" Kiev, USSR (1972).
66. Variational Methods and Gravitational Waves; Ondes et Radiations Gravitationelles, Colloque Internationaux C.N.R.S. No. 220 Paris 18–22 Juin 1973, pp. 57–71.
67. Gravitational waves and averaged Lagrangians, Proceedings of Symposia in Pure Mathematics Vol. 27, 415–423 (1975).
68. High frequency gravitational radiation in Kerr–Schild space-times, *Comm. Math. Phys.* **47**, 185–196 (1976).
69. Curvature Invariants, Characteristic Classes, and the Petrov Classification of Space-times. In "Differential Geometry and Relativity," M. Flato and M. Cahen, eds. Reidel, Dordrecht-Holland, 1976, pp. 277–289.
70. Lanczos' splitting of the Reimann tensor, *Comp. & Maths. with Appls.* **1**, 377–380 (1975).
71. Spatially homogeneous universes, *Proceedings of Marcel Grossman meeting on the Recent Progress of the Fundamentals of General Relativity*, 7–12 July 1975, Trieste, Italy, pp. 231–242.
72. High frequency, self-gravitating, charged scalar fields (with Y. Choquet-Bruhat), *Gen. Relativity Gravitation* **8**, 561–571 (1977).
73. Relativistic fluid mechanics, *Annual Review of Fluid Mechanics* **10**, 301–332 (1978).
74. High frequency gravitational waves, two-timing, and averaged Lagrangians. In "General Relativity and Gravitation," vol. 1, A. Held, ed. Plenum, New York, 1980.

Books Edited by A. H. Taub

In the "Collected Works of John von Neumann:"

Vol. I, "Logic, Theory of Sets, Quantum Mechanics." Pergamon Press, Oxford, 654 pp. 1961.
Vol. II, "Operators. Ergodic Theory and Almost Periodic Functions in a Group." Pergamon Press, Oxford, 565 pp. 1961.
Vol. III, "Rings of Operators." Pergamon Press, Oxford, 574 pp. 1961.
Vol. IV, "Continuous Geometry and Other Topics." Pergamon Press, Oxford, 516 pp. 1962.
Vol. V, "Design of Computers, Theory of Automata, and Numerical Analysis." Pergamon Press, Oxford, 784 pp. 1963.
Vol. VI, "Theory of Games, Astrophysics, Hydrodynamics and Meteorology." Pergamon Press, Oxford, 538 pp. 1963.

"Computers and Their Role in the Physical Sciences" (with S. Fernbach). Gordon and Breach, New York (1970).
"Study in Applied Mathematics," Vol. 7 in Mathematical Association of America Study. Prentice Hall, Englewood Cliffs, New Jersey, 1971.